ALBUQUERQUE

*A Guide to Its Geology
and Culture*

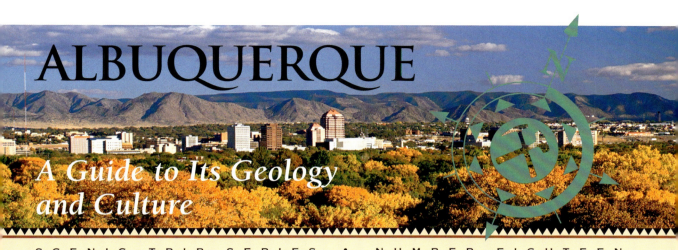

ALBUQUERQUE

A Guide to Its Geology and Culture

SCENIC TRIP SERIES • NUMBER EIGHTEEN

by
Paul W. Bauer
Richard P. Lozinsky
Carol J. Condie
L. Greer Price

New Mexico Bureau of Geology and Mineral Resources
A division of New Mexico Institute of Mining and Technology
Socorro, New Mexico
2003

Copyright © 2003 New Mexico Bureau of Geology and Mineral Resources
Peter A. Scholle, *Director and State Geologist*

A Division of New Mexico Institute of Mining and Technology
Daniel H. López, *President*

801 Leroy Place
Socorro, NM 87801- 4796
(505) 835-5420
http://geoinfo.nmt.edu

DESIGN & LAYOUT: Christina Watkins, Amanda Summers
CARTOGRAPHY & GRAPHICS: Kathryn Glesener, Rebecca Taylor, Thomas Kaus, Leo Gabaldon.
EDITING: Jeanne Deardorff, Nancy Gilson, Jane C. Love
SERIES EDITOR: L. Greer Price
COVER PHOTOGRAPHS: Bosque with Sandias in distance, George H. H. Huey; inset of Albuquerque skyline, Marble Street Studios (inset)
Back cover: Downtown Albuquerque ca. 1881, at the corner of what is now Central Avenue and First Street, courtesy of Albuquerque Museum

Library of Congress Cataloging-in-Publication Data

Albuquerque : a guide to its geology and culture / by Paul W. Bauer ... [et al.].-- 1st ed.
p. cm. -- (Scenic trip, ISSN 0548-5983 ; no. 18)
Includes bibliographical references and index.
ISBN 1-883905-14-1
1. Geology--New Mexico--Albuquerque Region--Guidebooks. 2. Albuquerque Region (N.M.)--Guidebooks. I. Bauer, Paul W., 1956- II. Series.
QE143.A3no. 18
[QE144.A]
557.89 s--dc21
[557.89/   2003001339

Scenic Trip 18
First edition 2003
Printed in Canada

DEDICATION TO VINCENT C. KELLEY

Vincent Kelley, author of the first versions of the Albuquerque Scenic Trip, was one of New Mexico's most productive and influential geologists. Educated at UCLA and Caltech, Vin arrived at the University of New Mexico in 1937 to assume a faculty position in the Department of Geology. His career on the faculty spanned 33 years (1937–1970), the last nine as chair of the department, followed by another 12 productive years of research and publishing as Professor Emeritus before his death in 1988 at the age of 84. Trained as a classical field and economic geologist, Vin combined a wealth of field observations into careful syntheses and detailed geologic maps of the structural and stratigraphic framework and history of many parts of New Mexico. Edward Beaumont, a former student and close friend, once remarked:

Vin Kelley's achievements as a field geologist are legendary. He possessed boundless energy, keen powers of observation, tremendous determination, but more than that: a total lack of fear. He went places and did things that others thought impossible or inordinately dangerous.

In these times of increasing specialization within the geosciences, geologists of Vin's type, widely knowledgeable in a broad spectrum of geologic fields, are all but extinct.

Among his more than 150 publications are many lengthy, comprehensive studies of entire mountain ranges (including the Caballo and Sandia Mountains), large structural/stratigraphic basins (including the Albuquerque and Española Basins), and other large areas of the state, including the Pecos country of southeastern New Mexico. In addition, he contributed much to our understanding of the tectonics of the Colorado Plateau, mineral deposits of New Mexico and Colorado, and the evolution of what has come to be called the Rio Grande rift. One would be hard pressed to suggest anyone who has contributed more to our knowledge of the geology of New Mexico. In the years that have passed, new generations of geologists have examined the rocks and structures that Vin studied, in some cases assembling more data and refining his conclusions. Yet Vin Kelley's work remains the firm foundation upon which new discoveries and interpretations have been constructed.

Vin trained hundreds of undergraduate geology students and nearly 70 masters and doctoral students at the University of New Mexico, many of whom achieved prominence in academic institutions, the U.S. Geological Survey, and industry. He was a teacher in the broadest sense—eager and happy to pass his knowledge on to others, instilling his own high scientific and professional standards and desire to learn from the rocks into several generations of students. His students regarded him with great respect, a little awe, and a high degree of devotion.

Vin ably documented the geology of the Albuquerque area in earlier versions of this book. More than two decades have passed since their initial publication, however, and new information and interpretations concerning the varied geological features of the Albuquerque region have been published by dozens of workers since then. The population of the Albuquerque urban area has grown by more than 60 percent since the early 1970s. This accelerating human impact upon the land and its resources, especially water, is an increasingly important part of the story of the geology of the Albuquerque area. It is fitting that this latest volume of Scenic Trips be dedicated to Vincent Kelley, who spent most of his professional life in Albuquerque, and who contributed so greatly to our understanding of the geology of the city and state.

Barry S. Kues
Department of Earth and
Planetary Sciences
University of New Mexico

Preface

The New Mexico Bureau of Geology and Mineral Resources is the geological survey for the state of New Mexico. A division of New Mexico Tech, we are responsible for scientific investigations of the geology and mineral resources of the state. In addition to our technical monographs, we publish maps, periodicals, and books for the general public, including our popular Scenic Trip series. Since 1953 these books have guided travelers through select areas of New Mexico, with descriptions of easily accessible, visually spectacular geologic and cultural features.

The Albuquerque area qualifies with honors, exhibiting a remarkably diverse medley. The Sandia Mountains are the eroded

The New Mexico Bureau of Geology and Mineral Resources building, on the campus of New Mexico Tech in Socorro, houses a small publication sales outlet.

remnants of mighty Precambrian mountain belts; the overlying sedimentary rocks preserve a stratigraphic record of over 300 million years of earth history. The great Rio Grande valley, its volcanoes and earthquakes remind us of a more recent restless Earth. The trips examine a human history that spans thousands of years, from the prehistoric inhabitants of Sandia Cave to the Pueblo Indians, Spanish settlers, American adventurers, colonists, and the latest influx of immigrants. We may at last be stressing the area's natural resources to their limits.

This volume is the first in our newly redesigned Scenic Trip series. We've increased page size to allow for more (and better) photos and graphics, and we are now producing them in full color. This particular volume replaces the beautifully written third edition of the now out-of-print Albuquerque Scenic Trip #9 by the late Dr. Vincent C. Kelley. We have attempted to maintain his same informal style, and we have borrowed heavily from his writing. However, Albuquerque and the surrounding areas have grown and changed rapidly in the last 15 years, and many new geologic discoveries and insights have been gained. We have added material on cultural history, fossils, the evolution of the Albuquerque Basin, the Turquoise Trail, and many other topics.

We have included six auto trips and one tram excursion in this book. The trips range in length from 20 to 80 miles. To fully appreciate the trips, and to allow time for your own spontaneous stops, we recommend a full day for each of them. Please respect private property, never disturb archaeological and historical sites, and do not remove any artifacts. If you are interested in additional information, check the Suggested Reading lists at the end of each chapter.

If you are unfamiliar with the use of road logs, read the introduction on page 67 before starting a trip. In an urbanized area like Albuquerque, safety is particularly important. Never stop in the middle of a road, and pull off only where there is room (and it is safe) to do so. Drive safely, tread lightly, and enjoy the scenery!

Paul W. Bauer
Richard P. Lozinsky
Carol J. Condie
L. Greer Price

Maps continue to be an important product at the bureau.

Table of Contents

INTRODUCTION 1

CHAPTER ONE
ALBUQUERQUE'S GEOGRAPHIC &
GEOLOGIC SETTING 5
    Geographic Setting
    Geologic Setting

CHAPTER TWO
A PRIMER ON ROCKS, TIME,
& TECTONICS 9
    Classifying Rocks
    Geologic Time and Rock Units
    The Rock Cycle
    Plate Tectonics

CHAPTER THREE
GEOLOGIC HISTORY OF THE
ALBUQUERQUE AREA 17
    The Rio Grande Rift
    Early Geologic Studies
    Earthquakes in Albuquerque
    The Albuquerque Volcanoes
    Earth Resources

CHAPTER FOUR
PRECIOUS WATER 39
    A History of Water Use in
      Albuquerque
    New Mexico Water Law
    The Middle Rio Grande Conservancy
      District

Vesicular basalt

CHAPTER FIVE
CULTURAL DEVELOPMENT OF THE ALBUQUERQUE AREA 51

    Paleo-Indians

    Desert Archaic

    The Anasazi

    Arrival of the Diné and Spanish

    Pueblo Rebellion

    Mexican and U.S. Rule

    The Railroad and Route 66

    Albuquerque Today

CHAPTER SIX
THE SCENIC TRIPS 67

    TRIP 1 — The Western Flank of the Sandia Mountains 68

    TRIP 2 — The Crest of the Sandias & Beyond 79

    TRIP 3 — The Turquoise Trail 100

    TRIP 4 — Along the Rio Grande 119

    TRIP 5 — The Albuquerque Volcanoes 135

    TRIP 6 — The Rio Grande Rift to the Colorado Plateau 142

Albuquerque Museums 164

Glossary 166

A Word About Maps 170

Index 171

Photo & Figure Credits 178

Acknowledgments 180

About the Authors 180

Geologic Column 182

Geologic Map of New Mexico .... inside back cover

INTRODUCTION

Albuquerque is located in a picturesque setting along the banks of the Rio Grande, nestled between the towering Sandia Mountains to the east, and the volcano-capped mesa to the west. The mountains, valleys, and volcanoes attest to the area's long and complex geologic history. If you understand their language, the rocks can recite tales of landscapes that have changed dramatically during the last 2 billion years, from shallow tropical seas to vast, sand-duned deserts; from enormous white sand beaches to meandering rivers and lush fern forests; from enormous mountain ranges to flat, featureless plains. These geographic features have also greatly influenced 10,000 years of human occupation. On the most basic level, prehistoric human inhabitants were surely awed by the dazzling sunsets, just as we are today. On a more practical level, the arrangement of Albuquerque's eclectic physiographic features—the bold mountain escarpments, the fertile, flood-prone valley, and the dry tablelands—has had a substantial impact on the anatomy and character of the city and its surrounding settlements.

With over 500,000 people, the Albuquerque urban area contains more than a third of New Mexico's population, most of which is concentrated along the Rio Grande. In 1891 the city covered only 2,000 acres; by 1983 it had grown to over 68,000 acres. The city has experienced tremendous population growth over the last 30 years. Between 1963 and 1981 the city increased by 40,968 single-family homes, 36,599 apartment units, and 3,945 commercial buildings. Between 1950 and 1980 the length of paved roads quadrupled to more than 1,700 miles, and in 1993 the city patched 25,331 potholes. Although the city has worked to attract high-technology industries, the predominant use of

OPPOSITE: **The Old Town plaza today.**

Central Avenue, downtown Albuquerque, ca. 1900.

the land in the scenic trip area continues to be agricultural. Along the Rio Grande floodplain, irrigated farms produce alfalfa, grass hay, corn, and chile. In the sparsely populated outlying areas, ranching still predominates. Within the Rio Grande valley, the sparse vegetation includes grasses, small shrubs, and succulents such as cacti and yuccas. Cottonwood, willow, and salt

Albuquerque in 1935. The Rio Grande and Barelas Bridge are visible in the lower left.

The same area in 1996. Note the increased channelization of the Rio Grande, and the thick bosque. Arroyos have been replaced by streets and diversion channels.

cedar dominate the Rio Grande floodplain. At higher elevations, grasses give way to piñon pine and juniper, which in turn are succeeded by the mixed conifer forests.

A combination of latitude and altitude accounts for the city's pleasant year-round climate. It is far enough south to avoid most of the cold northern storms, and, at 5,000 feet in elevation, high enough to avoid the heat that plagues cities of the same latitude but lower elevations such as Phoenix, Arizona. The four distinct seasons that Albuquerque enjoys are also due to its elevation. Furthermore, the Sandia Mountains provide a blockade to outbreaks of frigid east-side air, humid summer air, and the tornadoes of the Great Plains. The climate of the Albuquerque area is warm, semiarid, with an annual precipitation of less than 10 inches that increases with elevation to over 20 inches in the Sandia Mountains. Typically, more than half of the precipitation falls in the late summer months as brief, intense, localized thunderstorms, fondly known as monsoons. The combination of steep slopes, low vegetation, and energetic storms provides a setting for spectacular short-term geologic events such as floods and debris flows. Add to that a major sprawling city with suburbs, and the result is a serious challenge for geologists and planners.

The articles and road logs in this guidebook are intended to acquaint the reader with the geologic and cultural history of the Albuquerque area. We have attempted to relate the two, and when we encountered an especially interesting geologic feature or human endeavor, we presented more in-depth discussions. We hope that this information adds to your enjoyment and appreciation of this spectacular region.

The museum at Coronado State Monument, Bernalillo.

The Rio Grande bosque.

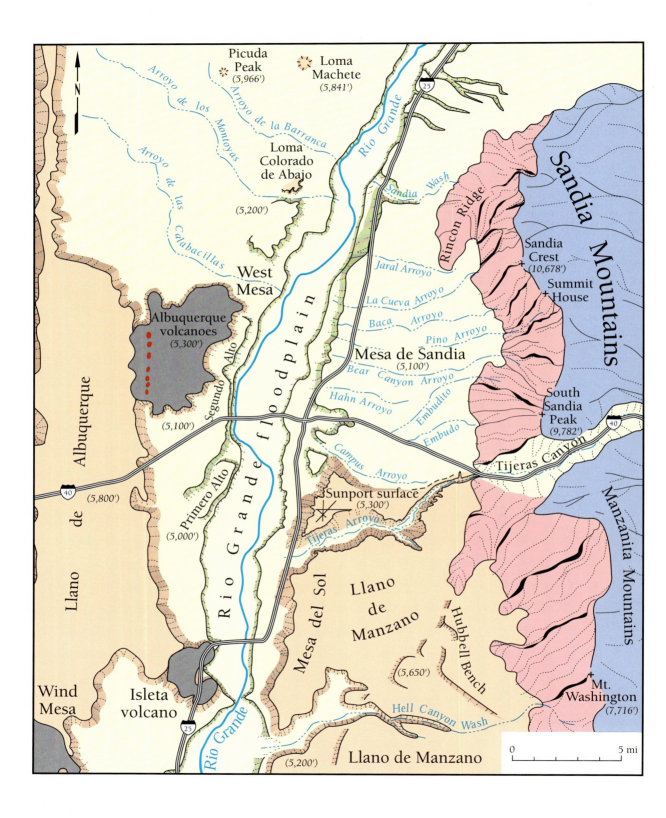

CHAPTER ONE

ALBUQUERQUE'S GEOGRAPHIC & GEOLOGIC SETTING

The major landscape feature in central New Mexico is the Rio Grande valley with its resplendent river, the Rio Grande (or, as it is known in old Mexico, the Rio Bravo del Norte). The Rio Grande is a major western river. From its headwaters high in the Rocky Mountains of southern Colorado, it flows southward 1,640 miles to the Gulf of Mexico. Remarkably, the river is nourished by only a handful of major perennial tributaries during its long journey. Near Albuquerque, the river flows along a 2.5 to 4-mile-wide floodplain that slopes gently southward at about 4 feet per mile. River bed elevation just west of downtown Albuquerque is 4,950 feet above sea level.

West of the floodplain, the land rises through low bluffs and gradual slopes to a high mesa 5,600–6,000 feet in elevation, known as the Llano de Albuquerque ("Plain of Albuquerque"), which narrows southward. The Rio Puerco, a major tributary of the Rio Grande, borders the Llano to the west until it joins the Rio Grande to the south.

Immediately east of the floodplain, the land rises gradually for about 8 miles to an elevation of about 6,000 feet at the base of the Sandia Mountains. From there, the slopes rise abruptly to the rugged crest of the range. The highest point in the mountains is Sandia Crest at 10,678 feet, more than a mile above the river. Just south of Albuquerque is Tijeras Arroyo, the major drainage of the Sandia Mountains. Tijeras Creek has eroded a pass through the mountains and a deep canyon from the mountains to its confluence with the Rio Grande. The back side of the Sandia Mountains slopes more gently eastward to merge with the plains of the Estancia Valley.

GEOLOGIC SETTING

THE ALBUQUERQUE BASIN To say Albuquerque is located in a valley is geologically imprecise. From central Colorado to El Paso, Texas, the river flows through a series of linked depressions or basins known collectively as the Rio Grande rift. These basins began forming about 30 million years ago when the earth's crust began stretching in an east-west direction. Crustal blocks within the center formed downdropped basins, whereas the adjacent blocks formed the topographically high bounding uplifts. Large normal faults separate the basins from the uplifts. Naturally, as the basins dropped and the mountains rose, sediments began to erode from the mountains to fill the basins. Most of the sediments were transported by rivers and streams. So, rather than excavating this enormous valley, the Rio Grande actually helped to fill a gigantic depression, reworking sediments that have entered the

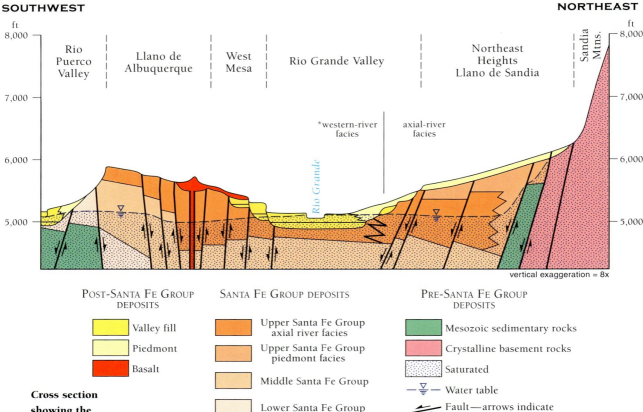

Cross section showing the subsurface geology of the Rio Grande rift in the Albuquerque area.

basin from many sources. It is geologically more precise to say that Albuquerque is located within a structural basin.

The Albuquerque Basin is one of the largest in the Rio Grande rift, measuring about 25 miles wide and 100 miles long. Deep oil-test wells drilled within the basin have shown that the sediments filling the basin are locally over 15,000 feet thick. These sediments, known as the Santa Fe Group, represent a combination of material eroded from the flanking uplifts and material transported into the basin, mainly by rivers and streams. The rift-basin structure is complex, generally consisting of a deep, inner basin flanked by relatively shallow benches (such as the Hubbell Bench just south of Albuquerque) that step up to the margins.

THE MOUNTAINS The Sandia, Manzanita, Manzano, and Los Pinos Mountains form the prominent eastern boundary of the Albuquerque Basin. These uplifts are giant blocks of crust tipped slightly eastward. The tipping is similar to the result of stepping on the edge of a loose brick, causing one side to rise and the other to sink. From Albuquerque, we have a view of the bold, upturned side. These mountain ranges consist of a crystalline core of Precambrian plutonic and metamorphic rocks unconformably overlain by Paleozoic limestone, shale, and sandstone. The Sandia and Manzano blocks tilt only about 15° eastward, but because of the simultaneous subsidence of the Albuquerque Basin, the spectacular, 14-mile-long western escarpment averages 4,000 feet in height. The eastern side of the mountains has a much smoother appearance because erosion has removed the stratified surface, layer by layer.

By calculating the vertical distance between the Precambrian/Paleozoic contact on Sandia Crest with the same contact buried deep in the basin, the amount of movement along the basin-bounding faults can be estimated. This distance is well over 30,000 feet. That is to say, the trough created by the formation of the Albuquerque Basin is 5–6 miles deep—one of the deepest such holes in the world!

The western basin boundary is not as well defined by physiographic features. The Sierra Ladrones and the Lucero uplift form the southwestern boundary. The Sierra Ladrones are composed of Precambrian granitic and metamorphic rocks, whereas Paleozoic limestone, sandstone, and shale capped by late Cenozoic basalt flows form the gently west-tilted Lucero uplift. North of the Lucero uplift, the topographically subdued Rio Puerco fault zone marks the boundary between the Rio Grande rift and the Colorado Plateau. Rocks exposed west of the fault zone include Cretaceous sandstone and shale.

It's a landscape of remarkable diversity. The Sandia Mountains clearly dominate the city, but the Rio Grande is the heart of the region. And between the crest of the Sandias on the east and the high mesas west of the city lie many different environments and ecosystems, including riparian, desert, and alpine. It is this unique combination of climate, landscape, and the life-giving waters of the Rio Grande that has drawn people to this region for over a thousand years. There are few places where geology has so clearly influenced the cultural heritage of a region. It's very likely that it will continue to do so.

SUGGESTED READING

Albuquerque Geology, Frank J. Pazzaglia and Spencer G. Lucas, editors. Guidebook to the 50th annual field conference, New Mexico Geological Society, 1999.

Visitors Guide to the Sandia Mountains, Southwest Natural and Cultural Heritage Association in cooperation with Cibola National Forest, 1994.

The Sandia Mountains, along the crest looking north. Note the eastern tilt (to the right) of the mountain block and the sedimentary strata overlying the massive granite of the lower slopes.

CHAPTER TWO

A PRIMER ON ROCKS, TIME, & TECTONICS

Any rock that you find will belong to one of the three major rock groups: igneous, sedimentary, or metamorphic. Each of these groups is subdivided according to mineralogy and texture. Texture refers to the size, shape, and arrangement of the grains and/or crystals that make up the rock.

Igneous rocks originate from the crystallization of molten rock, called magma. Individual minerals in igneous rocks generally interlock with each other in all directions. Magmas that erupt onto Earth's surface produce extrusive (volcanic) rocks. They cool so rapidly that crystals do not have time to grow large. Thus, the texture of extrusive rocks is fine grained, and crystals are difficult to see without magnification. If magma is supercooled upon reaching the surface, the resulting texture is glassy and the rock is known as obsidian. Sometimes, during a volcanic eruption, the magma becomes frothy because of escaping gases. When this rock cools rapidly, some of this escaping gas is trapped in cavities forming pumice, a rock so light in weight, that it will float.

Alternatively, magmas that crystallize below the surface produce intrusive (plutonic) rocks. They cool slowly and therefore display minerals that are easily visible to the naked eye. Sometimes magma will initially cool slowly, then more rapidly. When this happens large crystals (phenocrysts), commonly quartz or feldspar, end up trapped in a mass of finer crystals, called groundmass. Such a texture is called porphyritic, and the rock is called a porphyry (e.g., granite porphyry).

In a very general way, igneous rocks with abundant iron and magnesium minerals (mafic minerals) such as biotite, amphibole, and pyroxene are darker in color than rocks with abundant potassium, sodium, aluminum, and silicon minerals (felsic minerals) such as quartz, feldspar, and muscovite. Igneous rocks can therefore be roughly classified by color and texture.

Sedimentary rocks fall into three groups: (1) those that have formed from the accumulation of rock and mineral fragments transported and deposited by water, air, ice, or gravity; (2) rocks that have precipitated from chemical solutions at Earth's surface; and (3) rocks that have formed from the remains of plants and animals. Within these three broad categories, sedimentary rocks are further classified by composition and texture. The

OPPOSITE:
The western face of the Sandia Mountains.

Andesite

CLASSIFICATION OF IGNEOUS ROCKS

TEXTURE	LIGHT QUARTZ	LIGHT NO QUARTZ	MEDIUM	DARK
Coarse-grained (plutonic)	granite	syenite	diorite	gabbro
Fine-grained (volcanic)	rhyolite	trachyte	andesite	basalt
Glassy	pumice	pumice	pumice	obsidian

fossilized remains of animals and plants may be preserved in any kind of sedimentary deposit. Consolidation by compression and cementation of grains turns sediments into sedimentary rocks, a process known as lithification.

Metamorphic rocks form in response to high pressures and temperatures encountered during deep burial. The process of metamorphism generally results in a new mineralogy, composition, texture, and/or internal crystal structure. Metamorphic rocks were never molten; instead they form as solids. Consequently, in some circumstances, their crystals grow in parallel alignment producing foliation. Regional metamorphism occurs over a large area during deep burial, whereas contact metamorphism occurs more locally at any depth when magma bakes the adjacent country rock. Metamorphic rocks are classified according to foliation, grain size, and mineralogy or color. Commonly, shale is metamorphosed to slate or schist, lime-

HOW DO GRANITES FORM?

A vast complex of granitic rocks are exposed in western North America; they range in age from Precambrian to Cretaceous. A belt of Precambrian granites is exposed in the Rocky Mountains from Canada to central New Mexico. Mostly younger granites (Mesozoic) are exposed in the North American Cordillera west of the Rocky Mountains, from Alaska to Mexico, and in the Andes of South America. Most granitic rocks are thought to result from the partial melting of rocks deep within the earth, from the lower crust or the upper mantle (at depths of about 30 miles). In plate tectonic framework, granites most commonly form at convergent plate boundaries where crust has been subducted, especially if the subduction occurred beneath continental crust. Such was the setting for most of the granites along the North and South American Cordillera. Granites may also form within extensional plate boundaries, such as in continental rifts. Granitic magma probably moves upward through the continental crust along zones of weakness. Some geologists believe that the most common shape for the intrusion is that of a mushroom, where the magma moves upward along narrow feeder dikes to spread out laterally, perhaps below a resistant rock layer. Such an igneous intrusion that is less than 5 miles in diameter is called a laccolith (Greek for "stone cistern"). Granitic batholiths are large areas of plutonic igneous rocks, commonly in mountain belts, that generally fall into the compositional range of diorite to granite. Batholiths are intrusions with exposed surface areas of more than 40 square miles and no known floors, whereas the term pluton refers to any igneous intrusion. Most batholiths consist of tens to hundreds of individual plutons of different sizes. Within a single batholith, individual plutons may exhibit widely different compositions, degree of deformation, and depth of emplacement.

No one knows how long it takes a pluton to solidify from a magma, as it happens out of sight and probably over thousands of years. Factors such as the size of the pluton, the temperature of the magma and the country rock, the presence of magmatic fluids, and the composition of the magma influence crystallization times.

PHYSICAL PROPERTIES OF SOME COMMON MINERALS

MINERAL	FORMULA	COMMON CRYSTAL FORM	COMMON COLOR	HARDNESS
Quartz	SiO_2	prism	colorless	7
Feldspar (orthoclase)	$KAlSi_3O_8$	blocky prism	white, gray, pink	6
Feldspar (plagioclase)	$(Na,Ca)Al_2Si_2O_8$	blocky tablets	colorless, white, gray	6
Mica (muscovite)	$KAl_2Si_3AlO_{10}(OH)_2$	hexagonal platelets	colorless	2½
Mica (biotite)	$K(Mg,Fe)_3Si_3AlO_{10}(OH)_2$	hexagonal platelets	black, dark brown	2½
Pyroxene (augite)	Ca,Na,Mg,Fe silicate	4- or 8-sided prisms	black	5–6
Amphibole (hornblende)	Na,Ca,Mg,Fe,Al silicate	long, 6-sided prisms	black, dark green	5–6
Garnet	Ca,Mg,Fe,Al silicate	faceted spheres	red, brown	6½–7½
Olivine	$(Mg,Fe)_2SiO_4$	granular masses	olive	6½–7

stone becomes marble, sandstone becomes quartzite, and granite turns to gneiss.

GEOLOGIC TIME & ROCK UNITS

The last 4 billion years of Earth history are subdivided on the geologic time scale into ages that represent specific intervals of time. The boundaries between these intervals generally represent major unconformities (gaps in the geologic record because of nondeposition or erosion) or other important geologic events. For example, the Mesozoic Era is known as the age of dinosaurs. Eras are further subdivided into periods, epochs, and ages. Every rock can then be assigned to some era, period, or epoch. The granites in the Sandia Mountains are Proterozoic in age, the limestones on top of the Sandia Mountains are Pennsylvanian in age, and the Albuquerque volcanoes are Pleistocene in age. We are currently living in Holocene time, the youngest epoch of the Cenozoic Era.

Geologists group rocks in the field into a hierarchy based on rock type, age, and environment of deposition. The basic mapping unit, a formation, is a sequence of rocks that has particular characteristics that distinguish it in the field. Each formation is usually named for geographic feature where it is best exposed or was first described (e.g., Sandia Formation for the Sandia Mountains). In some cases, the dominant rock type has been used to name the formation (e.g., Dakota Sandstone). A group is the next higher rank, and it consists of two or more formations that are lithologically related (e.g., the Madera Group).

THE ROCK CYCLE

The rocks and sediments that are exposed at the surface of Earth change with time. These changes include size, shape, chemical composition, and position. Material that is weathered and eroded from highlands by water, ice, and wind is

Quartz

CLASSIFICATION OF SEDIMENTARY ROCKS

ORIGIN OF SEDIMENTS	ROCK NAME	COMPOSITION	TEXTURE
Mechanical (disintegration)	conglomerate sandstone	gravel, cobbles, boulders sand	coarse-grained (over 2 mm) medium-grained ($\frac{1}{16}$ mm–2 mm)
Mechanical & chemical (disintegration)	shale or mudstone	silt, clay	very fine grained (< $\frac{1}{16}$ mm)
Chemical (minerals precipitated from water)	limestone dolostone gypsum rock salt	calcium carbonate magnesium/calcium carbonate calcium sulfate sodium chloride	very fine to coarse-grained very fine to coarse-grained fine- to coarse-grained fine- to coarse-grained
Organic (remains of plants and animals)	limestone diatomite coal	calcium-rich remains silica-rich remains plant remains	fine- to coarse-grained fine-grained noncrystalline

later deposited elsewhere, commonly in lowlands or under water. These eroded sedimentary materials may eventually become new rocks by the process of lithification, or they may be transported elsewhere. In time, some of these materials may be deeply buried, whereupon they can evolve into igneous or metamorphic rocks. This natural recycling process of breakdown, transformation, and reconstruction of the earth's crust is known as the rock cycle. Although the processes within the rock cycle may take many millions of years, such as the gradual erosion of a mountain or the growth of a coral reef, catastrophic events occur over a time span of hours or even minutes. Examples of such "natural disasters" include volcanic eruptions, tsunamis (tidal waves), flash floods, debris flows, and earthquakes. These processes have the potential to move large volumes of rocks and sediments in a short time. Geologists commonly find ancient examples of such events preserved in the rock record.

PLATE TECTONICS

Besides time, there is a second very important geologic variable. That variable is position on the surface of the globe. During Pennsylvanian time, about 300 million years ago, a great diversity of marine invertebrates were living in a warm, shallow sea. Seventy-five million years later, during Triassic time, tropical plants and animals lived along lush rivers in a warm and moist climate. In both cases, when these rocks were deposited, New Mexico was very near to the equator. About 66 million years ago, at the end of Cretaceous time, as dinosaurs were becoming extinct, New Mexico was at nearly

A PRIMER ON ROCKS, TIME, & TECTONICS

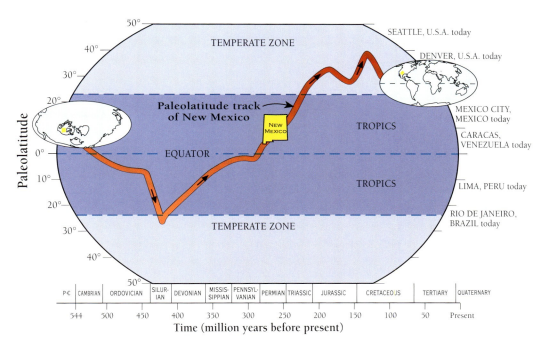

In the past 550 million years the North American continent (and New Mexico with it) has wandered from north of the equator to south, and back again. The climatic changes we see reflected in the geologic record are in part due to this global wandering.

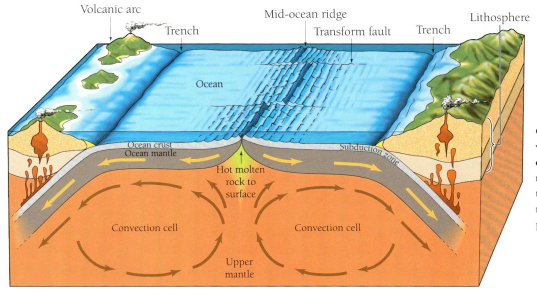

Convection within the earth's upper mantle is thought to be responsible for plate movement.

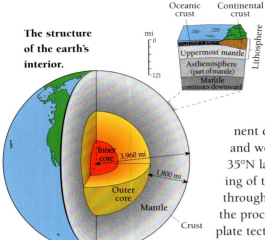

The structure of the earth's interior.

Geologists recognize three major kinds of plate boundaries.

50°N latitude, the present position of the Canadian border. Since then, the continent drifted southward, and we are now at about 35°N latitude. This drifting of the continents through time is part of the process known as plate tectonics.

According to plate tectonic theory, the uppermost layer of our planet, known as the lithosphere, is fractured into about a dozen rigid plates. These lithospheric plates move over an underlying layer of partially molten rocks known as the asthenosphere. Why do the plates need to move? The answer is that the earth is thermally stratified (hot in core, and cool at surface) and partially molten, so that hot, deep material flows upward where it cools and descends. These convection currents force the cool, rigid lithosphere to fracture, drift, and occasionally collide.

Most of the plate-related activity occurs at plate boundaries. Three general types of plate boundaries exist. Where boundaries converge, huge mountain systems develop (Himalayas, Alps, Andes) or great arcs of volcanoes form (Aleutian Islands, Indonesia). Where plates move past one another, great systems of steep faults form (San Andreas fault). Large magnitude earthquakes commonly occur at plate boundaries. Where plates diverge, new crust is created to fill the vacated space (Red Sea, East African rift). In some places, divergent boundaries begin to form within a plate. This may eventually lead to creation of two separate plates. The Rio Grande rift, which extends 600 miles through the center of New Mexico into Colorado, is an example of this. Along the Rio Grande rift in the Albuquerque area extension has pulled apart portions of the North American plate, extending the crust

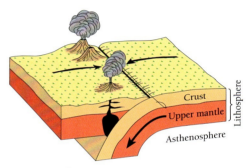
CONVERGENT PLATES

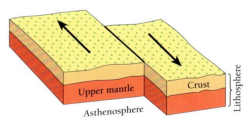
PLATES MOVING PAST EACH OTHER

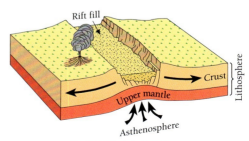
DIVERGENT PLATES

12-18 percent. This represents several miles of additional real estate over a period of 30 million years.

These are the basics that will help you understand the regional geology of the Albuquerque area. In examining the details, don't forget to look at them in the context of the big picture. The names are less important than the processes, and the processes responsible for the evolution of this landscape are the very same processes at work today. Last, but not least, in unraveling the geologic history of the area, keep in mind that events happening in other parts of the world, far removed from here, exerted an important influence on what was taking place here.

SUGGESTED READING

Basin and Range by John McPhee. Farrar Straus & Giroux, 1981.
A Short History of Planet Earth by J.D. Macdougall. John Wiley & Sons, 1996.
Beneath Our Feet: The Rocks of Planet Earth by Ron Vernon, Cambridge University Press, 2000.

CHAPTER THREE

GEOLOGIC HISTORY OF THE ALBUQUERQUE AREA

PRECAMBRIAN (3.8 BILLION – 544 MILLION YEARS AGO)

The North American continent consists of many distinctive pieces of crust, known as terranes, accreted (added) to the continent through geologic time by plate tectonic processes. In general, from north to south, the ages of these terranes range from the very old Archean (3.8–2.5 billion-year-old) rocks of Canada, Montana, and northern Wyoming to the Proterozoic (2.5 billion years–544 million years) rocks from Wyoming south to Mexico. New Mexico contains several distinct Proterozoic provinces, one of which is exposed in the Sandia Mountains. This province consists of rocks that range in age from about 1.7 to 1.4 billion years. Most of the Sandia Mountains is composed of granite and metamorphic rocks, visible from Albuquerque along the west face of the range. Precambrian rocks are also visible in Tijeras Canyon, on Rincon Ridge, and along the northern end of the range near Placitas.

Because of the great age and complex geologic history of these rocks, geologists have been able to piece together only a part of the Precambrian history. The oldest unit may be the Tijeras greenstone, which probably represents eruption of volcanic rocks such as basalt. At some later time, a large thickness of sedimentary rocks accumulated within a shallow marine basin. Over time, these layered rocks were moved deep within the earth by tectonic forces, where, approximately 1.4 billion years ago, they were metamorphosed and intruded by great bodies of molten rock that solidified into the Sandia granite. This granite, which forms most of the Sandia Mountain escarpment east of Albuquerque, is composed of large crystals of potassium feldspar, plagioclase feldspar, quartz, and biotite.

Unfortunately, we know little of the geologic happenings in central New Mexico for the next billion years, as a great unconformity separates the Precambrian granites from the Paleozoic sedimentary rocks above. This unconformity is exposed near the crest of the Sandia Mountains. This Great Unconformity, as it is often called, is similar in scope and age to the Great Unconformity exposed near the bottom of the Grand Canyon of the Colorado River, in Arizona.

PALEOZOIC ERA (544–245 MILLION YEARS AGO)

No rocks from the early Paleozoic remain in the scenic trip area, but geologists

> *The Precambrian has attracted geologists of exceptional imagination, who see families of mountains in folded schists.*
> —JOHN MCPHEE, *BASIN AND RANGE*

Fossiliferous Pennsylvanian limestone.

OPPOSITE: *Camarasaurus* mother and young feeding along the banks of a Late Jurassic stream, 150 million years ago. The remains of these giant sauropods have been found in beds of the Morrison Formation in the Albuquerque area.

Paleogeographic map of the Southwest during the Early Pennsylvanian. In the Albuquerque area, highlands shed erosional debris to form the terrestrial deposits we know as the Sandia Formation.

have been able to infer some of the early Paleozoic history of the Albuquerque area by studying excellent exposures in southern New Mexico. During Cambrian time (544–505 million years ago) northern New Mexico was a low-relief island that shed sandy sediments into shallow continental seas that nurtured trilobites and brachiopods. During Ordovician time (505–440 million years ago) most of the state was probably covered by a warm, shallow sea that teemed with early animals, including trilobites, bryozoans, brachiopods, snails, clams, corals, and 15-foot-long cephalopods (ancient relatives of modern squid and octopi). Later erosion stripped all evidence of Ordovician life from central and northern New Mexico. Similarly, central and northern New Mexico were stripped of all their Silurian (440–410 million years ago) marine sediments. During Early and Middle Devonian time (410–374 million years ago) most of the state was a low-relief, highly weathered landmass. In the Late Devonian (374–360 million years ago), much of the weathered rind of central New Mexico was eroded southward to accumulate as fossil-rich shales in the seas of southern New Mexico.

The oldest Paleozoic rocks in the Albuquerque area are Mississippian in age (360–325 million years ago). Although Mississippian limestones and shales blanketed much of the state, the Albuquerque area contains only patches of marine limestone (the Arroyo Peñasco Group) deposited in topographic lows on eroded Precambrian rocks. During Late Mississippian time, northern New Mexico was above sea level, and extensive caves developed in the porous limestones.

By Early Pennsylvanian time (300 million years ago), the major mountain building associated with the Ancestral Rocky Mountain orogeny had created a pattern of uplifts and intervening basins in central and northern New Mexico. The major

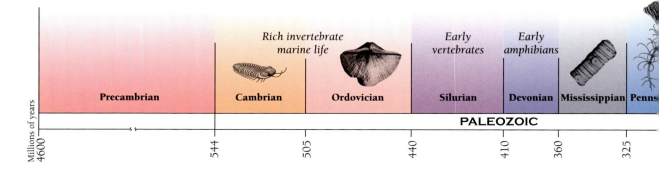

mountain belts present at that time included the Uncompahgre uplift of southwest Colorado, and the Sangre de Cristo uplift (the Ancestral Rocky Mountains) and Pedernal Highlands to the north and east. As these mountains rose, streams carried tremendous loads of erosional debris into valleys and marine basins. This debris formed the sandstones, shales, mudstones, and conglomerates of the Sandia Formation. In the Albuquerque area, the Sandia Formation averages about 200 feet thick and contains fossil wood and plant impressions of ferns and conifers. Over time, however, vast continental seas invaded the lowlands. Great thicknesses of marine rocks accumulated in the basins: 1,400 feet in the Tijeras Canyon area, and as much as 3,000 feet near the Ladron Mountains to the south. This complex, limestone-dominated sequence of Middle and Upper Pennsylvanian rocks, known as the Madera Group, contains abundant evidence of the varied ocean life of New Mexico 300 million years ago. The more than 300 reported species of fossil invertebrates from this area include brachiopods, corals, bryozoans, pelecypods, gastropods, crinoids, and the abundant small-shelled protozoans named fusulinids. Trilobites, which were common in early Paleozoic seas but rare in the Pennsylvanian fossil record, are found in abundance near Tijeras Canyon. From these fossil assemblages, paleontologists have deduced that the creatures lived in an open marine environment, in moderately deep water, far from any shoreline. Interestingly, although the swamps and marshes of the Pennsylvanian Period produced the greatest coal deposits in Earth's history, New Mexico was nearly barren of coal during this period due to its depositional setting. Instead, the landscape contained abundant cockroaches, giant dragonflies, spiders, tree ferns, scale trees, horsetail rushes, primitive conifers, and a few salamander-size amphibians.

In Permian time (286–245 million years ago), renewed mountain building to the north had forced most of central New Mexico to re-emerge from the Pennsylvanian seas. The youngest part of the Madera Group, the Bursum Formation, consists mainly of red sandstone and shale that represent the last phase of the Pennsylvanian/Early Permian marine environment. Rivers that flowed southward across the Albuquerque area built floodplains during the Early Permian (280 million years ago). Early land plants, giant amphibians, and primitive reptiles lived on

The division of geologic time into eras and periods is somewhat arbitrary, but it reflects some key events in the history of our planet. Although the earliest remains of life are found in the Precambrian, the fossil record is sparse until the beginning of the Paleozoic Era. Massive extinctions punctuate the record at the end of the Permian and the Cretaceous.

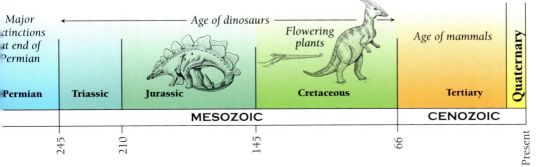

these floodplains. Impressions of leaves (especially of seed ferns) and the footprints of the amphibians and reptiles are the most common fossils of this time. Great volumes of highly weathered red sand and clay washed southward, driving back the shallow seas and their marine sediments. In the Albuquerque area, these rocks are crossbedded, locally feldspar-rich conglomerate, sandstone, siltstone, and shale of the 1,000-foot-thick Abo Formation.

Paleogeographic map of the Southwest at the end of the Jurassic. At this time most of New Mexico was exposed to erosion; deposits accumulated mainly in streams and small, closed basins.

The northern mountains had eroded sufficiently toward the end of Early Permian time to allow a shallow sea to advance and retreat repeatedly over most of New Mexico. The resulting pattern of alternating layers of marine dolomitic limestone, gypsum, and sandstone is preserved as the Yeso Formation. Late in the depositional history of the Yeso Formation, an extensive sheet of pure, white quartz sand flooded the Permian seas. The resulting Glorieta Sandstone is about 200 feet thick at its type section on Glorieta Mesa east of Santa Fe.

As the Permian seas continued to spread and deepen, yet another limestone was added to the Paleozoic layer cake. The San Andres Formation is as much as 1,000 feet thick in southern New Mexico, yet it is only 75 feet thick in the northern Sandia Mountains area. By late San Andres time, the seas had retreated to southern New Mexico. As the Permian ended, most of New Mexico rose above sea level, and the Paleozoic strata were slightly folded, warped, and eroded before deposition of rocks of Triassic age. The Permian/Triassic boundary also marks a great extinction episode in the history of Earth, as most species of vertebrates and invertebrates disappeared forever. However, the beginning of the Mesozoic also marked the beginning of a renaissance of life on Earth, the explosion of land animals, including mammals, birds, and, most dramatically, the dinosaurs. Exposures of Paleozoic rocks are seen along Trips 1 and 2.

MESOZOIC ERA (245–66 MILLION YEARS AGO)

The Albuquerque area had emerged from the sea by the end of Permian time, where it remained high and dry for the Triassic and Jurassic periods. Erosion was probably dominant from the latest Permian to early in the Triassic because no rocks of this age exist in the area. During the Middle Triassic, however, rivers from the north began depositing sands and muds of the Moenkopi Formation and Chinle Group. Warm and dry conditions prevailed into the Jurassic as indicated by the crossbedded, wind-blown sand of the Entrada

Sandstone and gypsum and limestone deposits of the Todilto Formation, deposited in a vast salt lake. Rivers again returned during the latter part of the Jurassic, depositing the multi-colored sands and muds of the Morrison Formation. The first record of dinosaurs in the Albuquerque area is preserved in Upper Jurassic strata of the Morrison Formation (about 150 million years old). Most spectacular are the giant sauropods *Camarasaurus* and *Seismosaurus*. Partial skeletons of these animals were excavated near San Ysidro during the 1980s. Rocks from the Triassic and Jurassic are exposed mainly in the northwestern portion of the scenic trip area (see Trip 5) where thicknesses range from 30 feet of Todilto Formation to over 500 feet for the Morrison Formation.

No sediments accumulated during Early Cretaceous time. By Late Cretaceous time, the shallow continental seas returned, depositing more than 100 feet of shoreline sands known as the Dakota Sandstone. As the sea continued to advance and deepen, muds of the Mancos Shale were deposited to thicknesses of over 500 feet. However, in contrast to the limey (calcium carbonate-containing) bottoms of the Pennsylvanian seas, the Cretaceous sea bottoms were muddy. Large, thin, plate-like clams called inoceramids inhabited the muddy bottoms. Coiled, extinct relatives of living squid and octopi, the ammonites, and sharks ruled the water column. Shells of inoceramids, and ammonites are among the most common fossils in the Mancos Shale. Fossil shark bones have not been found, because sharks have cartilage instead of bone, but shark teeth are found in abundance.

During latest Cretaceous time, the sea retreated and advanced several times, leaving behind a smorgasbord of sediments within the Mesaverde Group, including marine shale, shoreline sand, coal beds, and river-floodplain sand and gravel deposits. As the shoreline fluctuated during the Late Cretaceous, coal swamps formed. Here, densely vegetated jungles were inhabited by duckbilled and horned dinosaurs. Impressions of flowering plant leaves and dinosaur bones are the common fossils in the Mesaverde Group rocks that represent these swamp deposits. At the end of the Cretaceous, tectonic forces within the crust caused regional uplift of the area, and the seas retreated for the final time. This mountain building episode is known as the Laramide orogeny. The Laramide orogeny continued into the early Tertiary. Like all orogenies, it was a time of uplift, folding, faulting, intrusion of igneous rocks, and volcanic activity. It culminated in the uplift of the Rocky Mountains, and was responsible for many of the prominent structural features in this part of the country.

The end of the Cretaceous was also marked by one of the most massive extinctions noted in the geologic record. There is considerable debate over the cause of this extinction. Many (but not all) geologists believe it coincided with the collision of a large comet or asteroid in the vicinity of Mexico's Yucatan peninsula. Whether that was the single cause or one of a series of triggers in a more complex scenario is still hotly debated. Whatever the cause, nearly half of the species on earth—plants and animals, marine and terrestrial—were affected.

CENOZOIC ERA (66 MILLION YEARS AGO TO PRESENT)

The sea never returned to the Albuquerque area after its last retreat during the Late Cretaceous, and erosion dominated the landscape during the earliest part of the Cenozoic. By the beginning of the Eocene, 55 million years ago, central New Mexico was an inland floodplain far from the nearest seashore, which was in the Texas Gulf coastal plain. The Eocene floodplain developed under a subtropical climate and was inhabited by soft-shelled turtles and crocodiles that lived in rivers and lakes surrounded by palm trees. Beagle-size *Hyracotherium* ("Eohippus"), the first horse, and the hippo-like extinct mammal *Coryphodon* left their fossils in the red beds deposited on the floodplains. As the Laramide orogeny continued, downwarping of the crust in Eocene time formed basins that were rapidly filled with sediments of the Galisteo and Baca Formations. These sediments were eroded from nearby highlands and transported by streams into the Galisteo–El Rito Basin in the north and Carthage–La Joya Basin in the south. Early horses and rhinoceros-like brontotheres drank from these ancient rivers. The basins continued to receive sediments into Oligocene time. However, by that time the streams provided mainly volcanic detritus from large volcanoes to the southwest (Mogollon–Datil volcanic field) and possibly from the northeast (Espinaso volcanic field near Cerrillos).

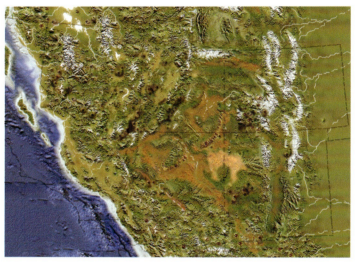

Paleogeographic map of the Southwest during the Oligocene. Depositional environments at this time consisted mainly of lakes and streams. The Rio Grande rift is just beginning to appear.

THE RIO GRANDE RIFT

About 30 million years ago, another major tectonic event began to reconfigure the New Mexican landscape. Due to plate tectonic events occurring a thousand miles to the west, at the boundary between the Pacific and North American plates, the New Mexican crust began to split apart. The initial development of the Rio Grande rift caused the scenic trip area to subside, forming the earliest Albuquerque Basin. The size of this basin is unclear, but by 10 million years ago the Sandia Mountains were rising rapidly enough to clearly define the eastern basin margin. Sediments, eroded mainly from the surrounding uplifts, were deposited into internally drained basins that contained small playa lakes (shallow lakes that contain water only during the rainy season). These earliest Santa Fe Group sediments are the Zia and Popotosa Formations. Evidence of volcanism within the basin during this early stage is evident at Black Butte south of Belen, a basalt flow at Trigo Canyon in the Manzano Mountains, and flows encountered by oil-test wells.

Increasing tectonic activity and basin development occurred between 10 and 5 million years ago when the flanking uplifts rose to nearly their present elevations and the sub-basins subsided rapidly. With the increase in tectonic activity, two major river systems, one flowing from the northwest and the other from the northeast, brought sediments into the basin. These river systems probably emptied into a large playa lake in the southern part of the basin. Volcanism increased during this time both within the basin (Hidden and Mohinas Mountains) and along the western basin margin (the Lucero uplift). About 15,000 feet of Santa Fe Group sediments accumulated during the interval from 30 million years ago to 5 million years ago. The Santa Fe Group has yielded a remarkable fauna of terrestrial vertebrate fossils, including camels, horses, antelopes, rhinoceroses, elephants, and small rodents.

A major change in depositional environment occurred about 5 million years ago when the Rio Grande became a through-flowing river system. Two other ancestral drainages, the Rio San Jose from the west and the Rio Puerco from the northwest, joined the Rio Grande to form a large plain of river sedimentation in the central Albuquerque Basin area. These rivers were broad, braided systems, quite different from the narrow, meandering rivers of today. The ancestral river sediments, together with debris shed from the surrounding uplifts, constitute the upper Santa Fe Group (Sierra Ladrones Formation and Ceja Member), which averages about 100 feet thick, but locally may be over 1,500 feet thick. (Exposures of the Santa Fe Group can be seen repeatedly on each of the road trips in this volume.) Volcanism continued both within the basin (e.g., Tomé Hill, Wind Mesa, Isleta, and Los Lunas volcanoes) and in the Lucero uplift. The three major ancestral rivers converged southward to form a single trunk system near Socorro. This ancestral Rio Grande flowed southward along the rift to eventually empty into a large closed basin in northern Chihuahua, Mexico.

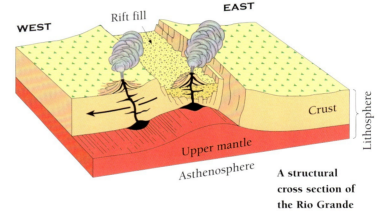

A structural cross section of the Rio Grande rift.

Another significant change in the depositional history occurred between about 1 million and 600,000 years ago when the Rio Grande began its first major downcutting episode. This may have been in response to integration with drainages to the Gulf of Mexico and receiving more runoff from the southern Rocky Mountains. This was a critical event in the history of the Albuquerque Basin because before this time the basin had been primarily filling with sediments. During the last 500,000 years, the river has established a history of general downcutting, as several more cycles of river

The first geologic map of New Mexico was published in 1858 by French geologist Jules Marcou, who accompanied the Whipple Expedition across the New Mexico Territory in 1853.

incision have produced the modern inner valley of the Rio Grande.

Tectonism within the basin has slowed during the last million years. Uplifts around the basin continued to shed detritus into the basin, but not as rapidly as before. The cycles of river erosion and aggradation (sediment accumulation) coincide with wet and dry periods of the Pleistocene ice age. Wet periods provided more water, causing downcutting, whereas dry periods resulted in partial backfilling of sediments, forming river terraces. Volcanism has continued with the eruption of the Albuquerque and Cat Hills volcanic fields on the Llano de Albuquerque. Today, uplift, subsidence, erosion, and deposition occur throughout the basin but at relatively low rates. In the last 10,000 years, the Rio Grande valley has accumulated about 80–100 feet of sediment.

The youngest fossils in the Albuquerque area come from gravel pits and arroyo cuts in Pleistocene river gravels and pond deposits, mostly within the Albuquerque city limits. The gravel pits commercially mined in the city have long produced teeth and bones of camels, mammoths, horses, and ground sloths. They represent a large-mammal fauna that became extinct in North America about 10,000 years ago, at the end of the last Ice Age. A 1.5-million-year-old pond deposit near the Albuquerque International Airport indicates a cottonwood bosque grew there under a much wetter and cooler climate than now. Fossils thus show that a visit to the Albuquerque area during the Pleistocene would have been like visiting a modern East African game park with a climate much like that of present-day Los Alamos.

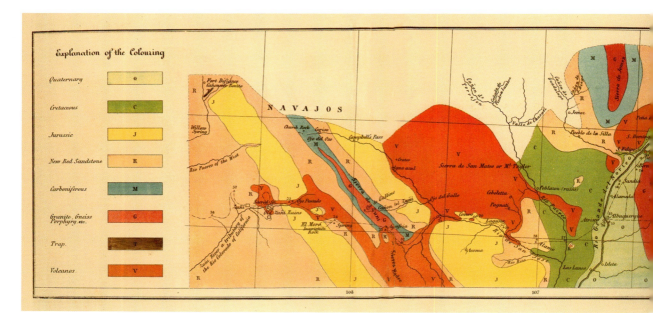

EARLY GEOLOGIC STUDIES

The earliest records of geologic observations in the area are found in the sixteenth century chronicles of early Spanish explorers who were mainly interested in gold, turquoise, copper, sulfur, and salt. In 1807 Lt. Zebulon M. Pike (U.S. Army) was arrested by Spanish authorities and taken to Mexico via the Camino Real. As he passed through the Rio Grande valley, he recorded some general observations on the local geology. In 1828, when placer (a surficial sedimentary deposit of a mineral or native metal in unusually high concentration) gold was discovered around the Ortiz Mountains, hordes of prospectors scoured the area. In 1833 rich vein deposits of gold were discovered in the nearby mountains, and another placer field was found in the San Pedro Mountains in 1839.

New Mexico's first geologist, Thomas Falconer, was a member of the Geological Society of London. A member of the 1841 Texas–Santa Fe expedition, he was arrested and his minerals and journals were seized, but his journey represents one of the earliest geologic forays into New Mexico. The famous Santa Fe Trail trader Josiah Gregg wrote of New Mexican mines in his classic 1844 book *Commerce of the Prairies*. Gregg described gold mines, salt deposits, selenite used in windows, and petrified trees near Cerrillos. Several explorers investigated the region's geology in 1846; Lt. J. W. Abert reported "shark teeth, fish bones, fragments of large ammonites, and pieces of inoceramus" along the Rio Puerco, the first New Mexico fossils to be illustrated in a publication.

As the railroads pushed westward, several geologists were sent through New Mexico to do reconnaissance mapping. In 1858 one of these mappers, a Swiss–

French geologist named Jules Marcou, published Geology of North America with a chapter titled "Geology of New Mexico." Marcou collected fossils in Tijeras Canyon, and his journal states:

> I started with my friend Dr. John Bigelow, the botanist of the expedition, to ascend the highest peak of the Sierra de Sandia… The ascent of one of the most elevated summits of the Rocky Mountains—which after all is not a very easy matter, considering the wilderness, the difficulty of the roads and the fear of the Apache Indians—was effected by Dr. Bigelow and myself the 10th of October 1853. We chose the most elevated point of the Sierra de Sandia seen from Albuquerque, which attains the height of 12,000 feet above the level of the sea.

Besides scaling the Sandias, Marcou produced a colored geologic strip map across New Mexico.

Serious geological and paleontological studies began with the expeditions of Prof. John Strong Newberry who accompanied Lt. J. C. Ives in 1857–1858 and Capt. J. N. Macomb in 1859. During the 1870s, various collections of minerals and fossils were made by parties of the U.S. Geological Survey under Lt. G. M. Wheeler (after whom Wheeler Peak is named). In 1884 the great geologist Capt. Clarence E. Dutton was sent to New Mexico to study the volcanic features of the Rio Puerco valley. He produced a geologic classic, Mount Taylor and the Zuni Plateau, from a single summer's field work. Upon visiting Albuquerque, Dutton wrote:

> In the immediate valley of the Rio Grande the climate is temperate in winter and insufferable in summer; higher up the summers are temperate and the winters barely sufferable… Even the sagebrush, the ashy bloom of the desert elsewhere, resents the scorching summer and refuses to stay, and the cacti, vengeful and repellant everywhere, here assume a still more cruel and misanthropic mien.

The turn of the century brought many prominent and prolific geologists to the Albuquerque area. Two of the early presidents of the University of New Mexico were geologists: Clarence Luther Herrick, first professor of geology and second president, and his successor, William George Tight. Many UNM alumni later gained fame, including Kirk Bryan, whose senior thesis was titled "Geology of the vicinity of Albuquerque." Bryan's accompanying geologic map of the area was printed in five colors by Rand, McNally & Co. of Chicago. Native son Bryan went on to get a Ph.D. at Yale in 1920 and joined the faculty at Harvard in 1926. His research continued in New Mexico, focusing mainly on the Rio Grande depression (now called the Rio Grande rift), where his careful field work led to profound insights into the interconnected nature of the valley, mountains, volcanoes, and river.

Perhaps the most renowned geologist to work in New Mexico was Nelson Horatio Darton (1865–1948), once cited as America's most distinguished field geologist. In a career that spanned 60 years, mostly with the U.S. Geological Survey (USGS), Darton made enormous contributions to a variety of fields of geology. When Darton was assigned to the USGS Hydrographic Branch in 1895, his principal duty was to appraise the water resources of the West. Clearly, he needed

Kirk Bryan (1888-1950). His work bridged the gap between geologists and archaeologists working in the Southwest. His better known studies are works on the geology of Folsom deposits in New Mexico and the geology of Chaco Canyon.

to map the stratigraphy and structure to understand the hydrology (the study of water on and under the earth's surface), and in fact, he is best remembered for his geologic and topographic maps of vast areas of the West, including a fine 1:500,000-scale geologic map of New Mexico published by the USGS in 1922. The success of this map led to his production of similar state maps for Arizona and Texas. Another notable New Mexico publication was "Red beds" and associated formations of New Mexico, with an outline of the geology of the state (1928). Among his many publications are studies of Carlsbad Caverns, the Grand Canyon, and the Dakotas. Throughout much of the West, it was these maps and reports that established the stratigraphic units that have provided the foundation for all subsequent work.

EARTHQUAKES IN ALBUQUERQUE

Dozens of strongly felt earthquakes have shaken New Mexico during the past 150 years, and there is abundant evidence in the form of ancient fault scarps that the Rio Grande rift and adjacent regions have hosted strong earthquakes during the past 10,000 years. The first earthquakes reported in the state occurred near Socorro in 1849. However, because a modern network of seismographic instruments was not installed until the 1960s, much of the record is based on anecdotal accounts found in old newspapers, correspondence, and other documents.

The first permanent seismograph in New Mexico was installed in the

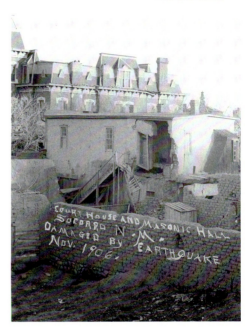

The November 1906 earthquake near Socorro (seen here) was felt in Albuquerque.
© J.E. Smith Collection

COMPARISON OF RICHTER MAGNITUDE AND MODIFIED MERCALLI INTENSITY

RICHTER MAGNITUDE	EXPECTED MODIFIED MERCALLI MAXIMUM INTENSITY (AT EPICENTER)	
2	I–II	Usually detected only by instruments
3	III	Felt indoors
4	IV–V	Felt by most people; slight damage
5	VI–VII	Felt by all; many frightened and run outdoors; damage minor to moderate
6	VII – VIII	Everybody runs outdoors; damage moderate to major
7	IX–X	Major damage
8+	X–XII	Total and major damage

From California Geological Survey

Socorro area in 1960 by New Mexico Tech, which has maintained an active research program in New Mexico earthquakes ever since. In 1961–62 the U.S. Coast and Geodetic Survey established a seismological station (now the USGS Albuquerque Seismological Laboratory) in the foothills of the Manzanita Mountains, southeast of the Albuquerque Sunport. Los Alamos National Laboratory set up a network of seismographs near Albuquerque in 1976.

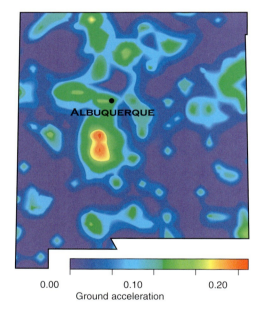

This probabilistic seismic hazard map shows the expected geographic distribution of ground motion from earthquake activity.

Early seismic stations were also installed in the state to monitor nuclear explosions (including two underground tests within the state of New Mexico in 1961 and 1967). Such studies have also enhanced greatly our understanding of the deep geology of the Rio Grande rift by measuring velocities at which seismic signals travel through the earth's crust.

Before the operation of permanent seismic stations, most earthquakes in New Mexico were reported from communities in the central Rio Grande rift (where most of the population has always lived), most notably between Albuquerque and Socorro. Since the 1960s, however, sensitive seismographic instruments have detected quakes throughout the state, including a magnitude 5 event in southeastern New Mexico near Eunice in January 1992. Many tremors in the central Rio Grande valley occur as part of extended swarms that last from days to many months, including those in Socorro (1849–1850, and again in 1904 and 1906), the Sabinal area (1893), and Belen (1935). Two earthquakes near Socorro in 1906, felt strongly in Santa Fe, Albuquerque, and El Paso, were the largest earthquakes in modern New Mexico history.

The intensity of an earthquake is a local measure of the amount of shaking felt and the damage done to buildings. Intensity is expressed using the Modified Mercalli intensity scale in units ranging from I (barely felt) to XII (complete devastation). Measured at many locations, intensity can provide clues to both how close and how big an earthquake is. In historical time, Albuquerque has experienced many mildly felt earthquakes, such as intensity III, and a few VII or greater. The common method used to describe the size of an earthquake is its instrumental magnitude. Magnitude is a measure of the absolute size of an event rather than its local effects, and is calculated using the amplitude of the wave produced on a seismogram.

Although most of Albuquerque's largest earthquakes occurred before 1962, and therefore before modern instrumentation was installed in the state, information can still be gleaned based on observed intensity and the area over which it was felt. For example, on May 28, 1918, one of New Mexico's strongest earthquakes shook the Cerrillos area. Based on effects such as chimneys and ceilings falling, the event was assigned a Mercalli intensity of VII or VII+ (considerable damage in poorly built or designed structures). These intensity measurements are consistent with a magnitude of about 5.5.

On December 3, 1930, Albuquerque experienced an intensity V–VI earthquake (some heavy furniture moved; instances of fallen plaster) with an estimated magnitude of 4. The New Mexico State Tribune (December 3 and 4) reported:

> At the University of New Mexico, the head of the English department, Professor George St. Clair, was reading poetry to his class on the second floor of Hodgin Hall. Having lived in the Philippines, he had experienced quakes before and stopped in the middle of a stanza to inform his class, "This is an earthquake!" He then proceeded to rush down the stairs, beating the entire class to the ground.

Earthquakes affect animals as well as humans. During the February 4, 1931, earthquake in Albuquerque (intensity VI+), happenings at the zoo were described as follows: "Pandemonium broke loose at the zoo. Animals awakened from their sleep, tore up and down their cages. A mad bedlam of noise issued from all parts of the park. The roars of the African lions were heard for a mile…." Another strongly felt earthquake shook the Albuquerque region (magnitude 4.5 and intensity VI) on November 6, 1947.

Albuquerque was the setting for intensity VI quakes on November 28, 1970 and January 4, 1971 (both magnitude 4.4), located near what is now the bustling west side. A summary of the 1970 event, taken from the U.S. Coast and Geodetic Survey earthquake report says: "Depth of 9 km (about 5.6 miles). Felt over about 3,100 km2 (1,200 mi2), principally in the Albuquerque region. Thousands were awakened in Albuquerque. Plaster cracked, windows broke, and many small items were broken. Roof of a barn collapsed. An air-conditioner on a roof shook loose and fell through a skylight. Other observers reported cracks in garage floor, exterior plaster cracks, and cracks in block fence walls. Many burglar alarms were activated. Animals were disturbed at the city zoo…." In December-January, 1997-1998 many Albuquerque residents felt earthquakes originating in the Willard area from a small swarm containing events up to magnitude 3.8.

Earthquakes in the Rio Grande rift area originate at quite shallow depths, ranging from about 3 to 6 miles. Most of this activity is associated with the sudden release of slowly accumulated stress along normal faults within and around the rift. The abnormally high historic level of earthquakes in the Belen–Socorro region, responsible for about half of the earthquake energy released in the state during the past 40 years, is attributed to the slow inflation of a shallow (12-mile deep) magma body underlying a portion of New Mexico larger than the state of Rhode Island.

Given the long history of moderate seismic activity and the prevalence of

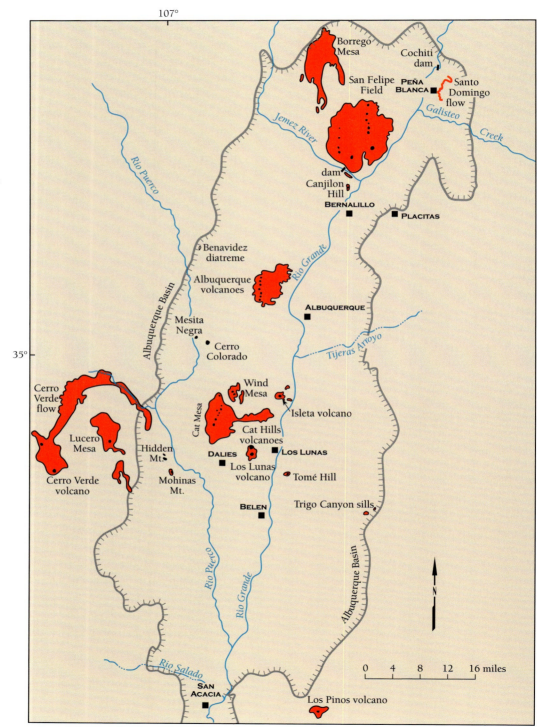

Volcanic features of the Albuquerque Basin.

adobe homes in the Albuquerque area, it is surprising that no deaths or serious injuries from earthquakes have ever been reported. The ancient fault scarps, however, attest to the region's long history of earthquake activity. One thing is certain: earthquakes will continue to shake the inhabitants of the Rio Grande valley.

THE ALBUQUERQUE VOLCANOES

Seven miles west of downtown Albuquerque lie the Albuquerque volcanoes, one of a series of volcanic fields in the Albuquerque Basin. The five large cones of the Albuquerque volcanoes form a distinctive (if subtle) western skyline to the city of Albuquerque. The volcanoes exposed at the surface are young, as young as 100,000 years old. The eruptions are concentrated along fissures, which parallel north/south-trending faults associated with the Rio Grande rift.

Initial eruptions began with the extrusion of very fluid lava along the fissures. Later stages of the eruptions involved more viscous (thicker) lavas that were centered on single vents. This pattern mimics eruptive patterns that we see today in active volcanic regions like Hawaii (see photo on page 139). This pulse of basaltic vulcanism was of short duration, possibly less than a decade. The rocks are primarily vesicular olivine basalt; small crystals of olivine and plagioclase may locally be seen with the naked eye. The vesicular texture formed as bubbles of gas, escaping from the ascending magma, were trapped in the rapidly solidifying basalt. The eruptions occurred on a gently sloping surface and the flows probably traveled slowly, incinerating all plant life but sparing animals that could flee at a rate of more than about 5-10 miles per hour.

Today the eroded lava flows can be seen up close at Petroglyph National Monument. The eastern edge of the flows (and many of the petroglyphs for which the region is famous) are best accessed from the east (Trip 4); the main visitor center is well worth a stop, and park staff can provide directions to the best sites. The three prominent volcanic cones visible on the horizon can be accessed from the west by way of the Volcanoes Day Use area (Trip 5). Hiking trails that begin at the parking area there offer an opportunity to hike into the volcanic cones.

In a state that boasts more volcanoes than any of the other lower 48 states, these are some of the youngest and most accessible. The southern Albuquerque Basin also contains the Cat Hills volcanoes, Isleta volcano, Wind Mesa, Los Lunas volcano, and the volcanoes of Canjilon Hill, Tomé Hill, and Mohinas and Hidden Mountains. Many of the volcanic features south of Albuquerque may be seen on Trip 6.

EARTH RESOURCES

In a state renowned for its mineral wealth, the Albuquerque area is better known for its green chile and sunsets than for its rich mines. With a few exceptions, only small or subeconomic mineral deposits have been discovered in the Sandia Mountains and immediate surroundings. These include some metallic minerals, nonmetallic rocks and minerals, and fossil fuels (coal and petroleum).

Before the arrival of the railroad to the

Albuquerque area in 1880, mining consisted of extracting those materials that were small and light enough to be traded (mostly precious metals) and those whose weight restricted them to local use only (pottery clay, adobe, stone, gypsum, mineral pigments, etc.). Although the Spaniards introduced mining to the area in the 1500s, systematic mining of precious metals did not begin until 1828 with discovery of gold in the Old Placers district of the Ortiz Mountains. After the American occupation in 1846, mining operations such as the Old Placers were modernized with steam engines and stamp mills.

With the arrival of the railroad, metallic ores and coal could be easily transported to distant smelters and markets, and lower grade deposits could be worked at a profit. Miners quickly switched from placer mining to the more lucrative lode mining. Although a few operations with especially rich deposits thrived, most outfits could afford to operate only during times of high metal prices.

The end of World War II offered another change to the mining industry of the area. Post-war inflation was accompanied by a doubling or tripling of the cost of labor, yet the price of gold remained fixed at pre-war levels. Although silver and base-metal prices rose, the increase in production costs outpaced them. The railroads switched from coal to diesel engines, and petroleum pipelines began to deliver cheap oil and gas to customers who had previously relied on coal. These factors caused the mining industry to shift toward the production of industrial minerals for the booming construction industry. Even today, the Albuquerque region is a major producer of sand, gravel, cement, and gypsum.

SANDIA MOUNTAINS AND TIJERAS CANYON Mining of metallic minerals in the Sandia Mountains area has been minor, although not for lack of trying. Small mines and prospect holes are common sights in the Precambrian and Pennsylvanian rocks of Tijeras Canyon and the Manzanita Mountains to the south. In Tijeras Canyon, small amounts of gold, silver, copper, fluorite, lead, and zinc were extracted from veins and fractures in mines such as The Great Combination, The Mary M, The York, and the Cerro

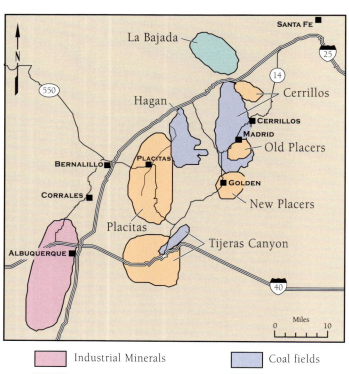

Mining districts in the Albuquerque area.

Pelon, all in Precambrian metamorphic rocks. Many of the world's great gold deposits are hosted in such rocks, on account of the tendency for metamorphicprocesses to concentrate and redistribute metal ores, but no mother lodes have been found near Albuquerque.

The best known mine in the Sandia Mountains is La Luz mine, which, at 10,040 feet in elevation, is one of the highest mines in the state. The La Luz Trail, which runs from the Juan Tabo Recreation area near the Sandia Peak Tramway up the front of the mountain to Sandia Crest, crosses near the mine at the head of La Cueva Canyon. Discovered in 1887 by Juan Nieto, a tunnel was driven into Sandia granite along a veined fault that contained lead and silver and some minor gold and copper.

PLACITAS DISTRICT The Placitas area contains many small prospects and mines, most of which produced little ore. The best known mine, the Montezuma mine, lies on the Las Huertas fault (which we cross on Trip 2) about 1 mile east of Placitas. Reportedly, a shipment of 21 tons of ore containing 12.5 percent lead and 11 ounces per ton silver was made in 1920. Deposits of barite, fluorite, quartz, and calcite plus minor metallic minerals on faults between Precambrian rock and Paleozoic limestone account for most of the mineralization in the Placitas area.

CERRILLOS DISTRICT The Cerrillos mining district encompasses 30 square miles of the Cerrillos Hills southwest of Santa Fe. The district is less notable for the value of past production than it is for its importance in the cultural development of the region. The turquoise deposits at Mt. Chalchihuitl are the most extensive prehistoric mining operations in North America. Archaeological and chemical evidence point to extensive mining and trading of Cerrillos turquoise across the continent for over 13 centuries. Artifacts from several prehistoric mines in Los Cerrillos show that Indians were also mining copper oxides for pigment and lead for pottery glaze. The Mount Chalchihuitl mine was excavated to a depth of 200 feet through solid rock by natives using stone tools in their search for turquoise. The area has dozens of prehistoric and Spanish Colonial mines, most small, some remarkably large given that they were dug with hand tools. In Spanish Colonial and Mexican time, from 1700–1849, vein deposits in Los Cerrillos were intermittently mined for sil-

Prospectors strike it rich, Cerrillos Mining District, ca. 1880. Henry Brown photo courtesy of Homer Milford.

Santa Fe Gold and Copper Mining Company smelter at San Pedro, ca. 1906. Henry Schmidt photo.

ver (bullion), lead (musket balls), and copper (armorplate and utensils).

The district did not boom until 1879, when most of the production was in zinc, lead, silver, copper, and gold. By 1890 miners were concentrating on the fabulous turquoise deposits, with millions of dollars worth of stone extracted by the end of the century. The early 1900s saw some metal production, but the boom was over. Turquoise production during the past 40–50 years has been negligible. As opposed to the Old and New Placer districts of the Ortiz and San Pedro Mountains to the south, the Cerrillos district never produced a significant amount of gold.

Los Cerrillos is underlain by a multiple intrusive complex of laccoliths, dikes, sills, and stocks (intrusives of less than 40 square miles in area) of Oligocene age that intrude Mesozoic and Eocene sedimentary rocks. There are two main intrusive lobes—the Bonanza lobe on the north, of which Turquoise Hill is an outlier, and the San Marcos intrusive center on the south that hosts the Cerrillos porphyry copper deposit and the Chalchihuitl turquoise. Turquoise occurrences are scattered throughout the district in patches of altered volcanic rocks. The turquoise occurs as nodules in open spaces and as irregular veinlets with little continuity. Within an individual deposit the color can range from sky-blue to apple-green.

NEW PLACERS (SAN PEDRO) DISTRICT
The New Placers district is located in the San Pedro Mountains near the town of Golden. The mines there have produced several million dollars worth of copper, gold, silver, lead, and zinc from rich veins related to contact metamorphism and metasomatism. The San Pedro mine, an

important mine in the district, dates from the mid-1800s. During the last 100 years, the mine has passed through many owners, and many smelters and mills have operated nearby. Also of note is the Golden placer, located in gravels along the northwestern part of the mountains. All large-scale placer operations in the district have failed, not because of the tenor of the gold, but because of the scarcity of water needed to separate gold from gravel.

OLD PLACERS (ORTIZ) DISTRICT Located in the Ortiz Mountains, the Old Placers district is one of the few 20th century mining success stories in the region. Known for placer gold deposits for a long time, it was not until 1978 that lode mining began at the Cunningham Hill mine by Gold Fields Mining Corporation. Much of the 250,000 ounces of gold recovered during the 8 years of mining corresponded to high gold prices, including the brief all-time high price of $850 per ounce. Although no mining has occurred during the past decade, other rich gold deposits have been identified in the mountains.

INDUSTRIAL MINERALS Nonmetallic mineral resources of the Scenic Trip area include collectible and unusual mineral specimens (such as barite, beryl, calcite, the silver mineral cerargyrite, cuprodescloizite or descloizite, garnet, graphite, specular hematite, ilmenite, ocher, pyromorphite, silver, strontianite, and tourmaline), building stone, rock aggregate, lime, shale and clay, and most important, cement ingredients such as limestone. The largest mine in the region is the limestone quarry of the Rio Grande Portland Cement Corp. (formerly the Ideal Cement Company) in Tijeras Canyon near the intersection of I–40 and NM–14/337. The plant cost $19,000,000 to build.

Crew at the Georgie shaft, Cerrillos Mining District, ca. 1880. Henry Brown photo courtesy of Homer Milford.

Production began in 1959 with the mining and milling of about 25,000 tons of limestone per year. Plant expansions have increased capacity to 500,000 tons per year. The high-calcium limestone beds quarried from the Madera Group average approximately 94 percent calcium carbonate, an excellent grade for the production of portland cement.

Other industrial minerals in the area include building stone, rock aggregate, lime, travertine, gypsum, and shale and clay. A variety of rocks from the Sandia Mountains have been used as building stones. Buildings, walls, and walks have been built from Precambrian granite, gneiss, quartzite, and schist, Pennsylvanian limestone, and sandstones of the Mesaverde, Dakota, Entrada, Santa Rosa, Yeso, Abo, and Sandia Formations. Mines along the Lucero uplift produce travertine used for building facing.

Loose rock aggregate consists of sand and gravel used mainly in construction and as a component of concrete. Abundant sand and gravel pits exist in the ancient Rio Grande river-related deposits of the Albuquerque Basin, and in the large alluvial fans along the base of the Sandia Mountains. At one time, lime (CaO) was produced from limestones of the Sandia Mountains, in kilns designed to burn limestone to drive off the carbon dioxide. The lime was then used in plaster, as a soil additive, in the smelting of ores, in refractory furnaces, and in many types of chemical processes. The only major shale/clay mining operation in the area is that of the Kinney Brick Company, located 9 miles south of Tijeras on NM–14. Clay-rich Pennsylvanian shales of the Madera Group have been quarried for brick-making material continuously since 1954. The clay is trucked to the Kinney brick plant in Albuquerque. The Albuquerque volcanoes have provided a local source of scoria, used mainly in landscaping.

HYDROCARBONS Although coal, oil, and natural gas have been discovered in the Albuquerque area, only coal has been successfully recovered. Nearly all of the coal comes from the Cretaceous strata of the Mesaverde Group, notably from the Hagan Basin north of the mountains, the Tijeras Basin, and in two spots west and northwest of Placitas. Coal has also been reported along the Rio Puerco in the western Albuquerque Basin and in many oil-test wells within the basin. Most of the Mesaverde Group coal seams are thin, discontinuous, and scattered throughout the formation. Because better sources of coal existed in the Gallup, Raton, and Cerrillos areas, commercial production of Sandia Mountains area coal never took off.

The Albuquerque Basin has a history of petroleum exploration that dates to the early 1900s. Nearly 50 oil and gas wells have penetrated the basin-fill sediments. The first was drilled in 1912 by the Tejon Oil and Development Company. Before 1950, exploration wells bottomed in the Santa Fe Group, at depths of less than about 6,500 feet. Several early wells intersected horizons that contained oil and gas, but none contained commercial quantities. As our understanding of regional geology increased, correlations with hydrocarbon-rich formations in the San Juan Basin of northwest New Mexico suggested that similar conditions should exist at depth in the Albuquerque Basin, and Shell Oil began a major exploration project of the basin in the early 1970s. Several Shell wells reported oil and gas shows (traces of oil or gas found in well cuttings) from Cretaceous rocks, but it was not until 1983 that significant quantities of gas were discovered in the West Mesa Federal No. 1 well. This discovery well was never put into production because of the "tightness" of the Cretaceous sedimentary rocks that contain the gas.

Hydrocarbon potential also exists within Cretaceous rocks buried in the Hagan Basin to the north and the Tijeras graben (a long depression bounded by high-angle faults) to the east of the Sandia Mountains. Sandstone lenses in the Mesaverde and Mancos strata might provide local stratigraphic traps, and large folds such as the Tijeras anticline might act as structural traps for oil and gas.

However, the Tijeras anticline (folded sedimentary rock that is convex upward) has been drilled several times without showing a trace of hydrocarbons.

SUGGESTED READING

New Mexico Rockhounding, A Guide to Minerals, Gemstones, and Fossils by Stephen M. Voynick. Mountain Press, 1997.

Volcanoes and Related Basalts of Albuquerque Basin, New Mexico, V.C. Kelley and A.M. Kudo. Circular 156, New Mexico Bureau of Geology and Mineral Resources, 1978 (reprinted 2001).

Geology of the Sandia Mountains and Vicinity, New Mexico by Vincent C. Kelley and Stuart A. Northrop. Memoir 29, New Mexico Bureau of Mines and Mineral Resources, 1975.

CHAPTER FOUR
PRECIOUS WATER

In the 1982 *Scenic Trip to Albuquerque*, Vincent C. Kelley stated:

> There is little or no chance of a water shortage, even with a substantial increase in population. Probably no other large city in the arid Southwest has such a bountiful supply of good water as Albuquerque. Water everywhere in the trough is no more than a few hundred feet beneath the surface.

Up until a few years ago, the city of Albuquerque had little reason to be concerned about its water supply. In 1982 UNM professor Vin Kelley confidently wrote that even if the city's population grew to 750,000 by 1990 (the city's population was only about 500,000 in 1990), and the city pumped 200,000 acre-feet of water per year (only 130,000 acre-feet were pumped in 1990), we would not empty the plumbing system for 170 years.

In fact, between 1989 and 1992 alone, water levels dropped 40 feet in eastern, northwestern, and south-central Albuquerque. In 1992 the New Mexico Bureau of Geology & Mineral Resources released a report titled "Hydrogeologic framework of the northern Albuquerque Basin" and the U.S. Geological Survey published a report titled "Geohydrologic framework and hydrologic conditions in the Albuquerque Basin, central New Mexico." The impact of these reports was immediate and dramatic; for the first time, the people of Albuquerque were told that their "inexhaustible" ground-water supply was rapidly disappearing. How could Vin Kelley and his colleagues have been so far off? The key lies in the enormous geologic complexities of the basin.

The Albuquerque Basin has abundant ground water because of a fortunate set of circumstances. For 25 million years, sand, gravel, silt, and clay of the Santa Fe Group have been filling the basin. Most of these deposits in the Albuquerque Basin are either (1) Rio Grande fluvial deposits (a sedimentary deposit composed of material transported by a stream) or (2) alluvial-fan deposits (unconsolidated sediment that fans out from the place where a stream issues from the mouth of a narrow mountain valley) or (3) piedmont-slope deposits

> *There is little or no chance of a water shortage, even with a substantial increase in population.*
> — KELLEY, 1982

OPPOSITE: New development on the banks of the Rio Grande between Rio Rancho and Bernalillo stand in stark contrast to the more traditional agricultural land uses on the far side of the river.

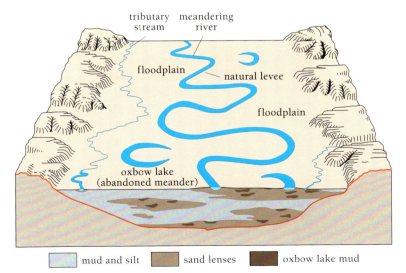

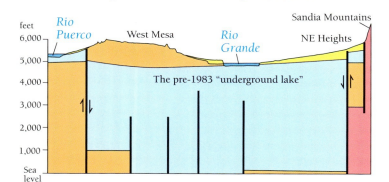

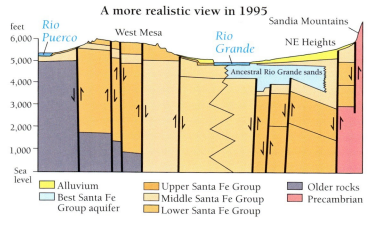

These two cross sections illustrate how advances in geologic understanding have altered our conception of the Albuquerque Basin.

(gentle sediment slope at the base of a mountain in a semiarid or desert region shed from the adjacent mountains). Sand and gravel deposits generally have considerable open spaces between the grains (porosity). It is within these pores that ground water is stored. A rock unit that stores and readily transmits water is known as an aquifer. Parts of the aquifer under Albuquerque are world class, in both storage capacity and water quality, but other parts we now know are not. The best aquifers are the youngest Rio Grande channel deposits. By circumstance, most of the city wells were drilled originally into the world class zones, and the hydrologists therefore assumed that most of the basin contained similar water wealth. However, we've discovered that most of the basin (and especially the deeper parts) consists of lower quality aquifers. Kelley had assumed that the good aquifers existed to a depth of at least 1,500 feet, but generally, the more deeply buried and older the sediments, the poorer the aquifer. As sediments become buried, their porosities and permeabilities decrease due to compaction and cementation, and water quality similarly deteriorates with depth.

Besides these vertical and horizontal variations in the sediments, the basin is cut by many large and small faults. Most of these faults are steep, normal faults that run north–south, parallel to the rift. Their effect on the distribution of aquifer zones is dramatic. They can place good aquifer zones against poor ones, so that closely spaced wells can have very different water yields. Many of these faults are not exposed or are hard to discern at the surface, and thus geologists and hydrologists must infer their locations from the exposed geology, geophysical studies, and well information. The net result of the combined complex stratigraphy and faulting is a huge three-dimensional jigsaw puzzle.

The surface that marks the top of the water-saturated zone is the water table. Water wells must penetrate below the water table to extract water from the pore spaces. The depth of the water table varies throughout the Albuquerque Basin. Along the Rio Grande floodplain, the water table is less than 10 feet below the surface; however, on the east and west mesas it is several hundred feet (and locally 1,000 feet) deep. The water table depth also fluctuates depending on the amount of recharge. Recharge here is primarily by surface water (from the Rio Grande and runoff from the mountains) percolating through the sediments. Geochemists have determined that some water in the aquifers may be ancient water that fell on the mountains thousands to tens of thousands of years ago, sometimes in much wetter climates. Today, we are in a dry period of geologic history, and consequently, recharge is low. Also, as the city has grown, surface runoff from the mountains and precipitation that falls on the city have been directed into a series of concrete-lined ditches and storm drains that move water directly into the Rio Grande. In the past, much of this water percolated into the ground, recharging the aquifer. So climate as well as growth and development are responsible for the fact that aquifer recharge is less than it has been in the past.

The most important factor influencing the level of the water table is the amount of water removed by pumping. Obviously, as people and industry move into Albuquerque, more water must be pumped out to serve them. In fact, the water level in some city wells has dropped as much as 140 feet in the last 30 years, and the trend will continue as long as ground-water withdrawals exceed recharge. This process of pumping more water than is being naturally recharged is called "mining water," and it can have serious consequences. If a critical volume of water is removed from enough porous sediment in an aquifer, the sediment may compress upon itself, decreasing the porosity and forever reducing the storage capacity of the aquifer. Such a process can also lead to ground subsidence at the surface and damage to buildings, utility pipes, and roads. Another consequence of dropping water levels is that it becomes more expensive to pump water to the surface. The long-term, tragic consequence of mining an aquifer is that ultimately the wells will go dry.

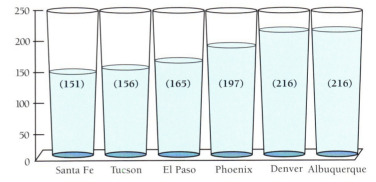

Water consumption for several southwestern cities in 2000 (in gallons per person per day).

The city is currently funding studies directed at determining the amount and quality of water remaining in the Albuquerque Basin aquifers. It turns out to be an extremely difficult problem because of the complex geology and the scarcity of data concerning subsurface conditions. Because the basin is partitioned into sub-basins by many faults, and because the character of the aquifer sediments commonly changes over short distances in all directions, it is difficult for geologists to thoroughly inventory the water resource with existing wells. The investigations continue with a new project that is drilling test wells placed strategically throughout the city.

Although the Albuquerque area has an adequate water supply for the near future, the aquifers are being depleted more rapidly than they can naturally recharge. The city's continuing growth means that greater and greater demands will be placed on the aquifer. An additional threat to the city's water supply comes from contamination of ground water by residential septic systems and industrial and petrochemical spills. It only takes a gallon of gasoline to contaminate a million gallons of water below drinking-water standards. A number of water wells in the city have been closed because of such contamination.

The city is currently working on ways to obtain more water, to increase recharge of the aquifers, and to curb contamination. The city has a standing offer to buy water rights from other water users along the Rio Grande. As developmental pressures make it unprofitable to farm in the Rio Grande valley, as farmland is subdivided and urbanized, and as the offering price for water rights increases, agricultural water is being transferred to the city. Ironically, it is the flood irrigation used in the valley that provides much of the recharge to the shallow aquifer. Among the recharge strategies that are being studied are injecting reclaimed water back underground, and using surface water from areas outside the Rio Grande drainage basin, such as water from the San Juan–Chama diversion project. Because such strategies are difficult and expensive, the city has also initiated a water conservation program whose goal is to reduce per person water use by 30 percent by the year 2004.

Water supply problems are not unique to Albuquerque. Nearby communities, several of which we will visit on the road logs, are concerned with water supplies and water quality. In Tijeras Canyon, the east mountain area, and Placitas, most of the serious problems are related to an enormous increase in domestic water use and septic systems. Ultimately, it may be water shortages that limit development in these areas.

A HISTORY OF WATER USE IN ALBUQUERQUE

Albuquerque is one of the oldest continuously occupied areas in the U.S. When the Spanish arrived in 1540, the Pueblo people already had many thousands of acres under cultivation in the middle Rio Grande valley. Under Spanish influence, elaborate networks of acequias (cooperatively owned and maintained irrigation supply ditches) were added; it is estimated that by 1800 over 100,000 acres of the valley were irrigated. Early settlers dug shallow wells by hand along the Rio Grande floodplain. This shallow ground water was more dependable and of higher quality than the river water itself. In the early 1900s, the city population grew slowly, and development was limited to the inner valley region where the water table was shallow. Although Willis T. Lee of the U.S. Geological Survey studied the water resources of the Rio Grande valley in 1907, it was not until 1936 that a comprehensive evaluation of water was conducted by the National Resources Committee.

The city's first well field was located near the present intersection of Tijeras and Broadway. Each of the nine hand-dug wells produced about 1,000 gallons per minute. One well reportedly measured 32 feet deep by 16 feet in diameter. This field was eventually replaced in the 1920s by the Main Plant well field, located along Broadway, approximately halfway between Central and Menaul. The 23 drilled (as opposed to hand-dug) wells of the Main Plant ranged in depth from approximately 200 to 500 feet. Shortly after that, the city began its expansion eastward out of the inner valley and onto the mesa.

Albuquerque's growth along the Route 66 corridor required new wells and water supplies, and the municipal water system was greatly expanded in 1949 with development of several well fields, including the Duranes field to the north, the San Jose field to the south, and the Atrisco field west of the river. Wells in these fields continue to produce water, although some have been abandoned or replaced.

During the 1950s the city experienced water supply problems due to declining water levels, and the system was expanded again beginning in 1959; included were the West Mesa field (where one 1,180 foot well produced 90° F water) and several fields in the Northeast Heights that were highly productive. The first Lomas field well was completed in 1962 near the base of the Sandia Mountains. Fields too close to the mountain tend to draw from thinner sequences of strata due to up-faulting and hence have lower production rates. Through the 1960s and 1970s additional wells were completed on the east and west mesas. Between 1954 and 1980, groundwater withdrawals tripled in the city wells.

Most of the city wells are completed in the Santa Fe Group, with the upper, dry portion of the hole cased off. The most permeable sediments in the Santa Fe Group, and most productive wells, are

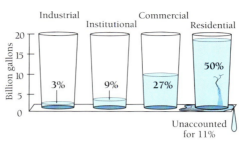

Billed water usage for the city of Albuquerque in 2000.

located about halfway between the mountain front and the inner Rio Grande valley, west of Eubank Boulevard. Early water-table maps from 1938 show that the water table sloped from northeast to southwest, roughly parallel to the Rio Grande.

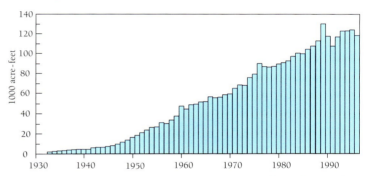

Annual ground water withdrawal for the city of Albuquerque (in thousands of acre-feet).

However, by 1960 water-table maps clearly showed the impact of increasing groundwater withdrawal in the Albuquerque area, as cones of depression (local areas of water levels lowered by pumping) had formed around several well fields. By 1978 the cones had become pronounced, especially under the Northeast Heights, reflecting the major growth of the city eastward. Locally, more than 130 feet of drawdown was measured in wells of the Lomas and Love fields. Between 1960 and 1978, it was estimated that 4 billion cubic yards of aquifer had been dewatered. The trend of groundwater consumption in the Albuquerque area is obvious: the first 300 years of development resulted in only minor impacts; however, well records from the last 60 years show that the impacts are large and are accelerating.

Albuquerque's municipal water system is supplied by a series of wells and reservoirs strategically distributed throughout the city. There are currently about 45 reservoirs fed by about 90 wells within the system, which is fully automated. From a master control building, all water flow and reservoir levels are recorded and controlled. Pumps in wells are automatically started and stopped as water is needed. Similarly, booster pumps that maintain adequate supplies and pressures to reservoirs are automatically controlled.

It is a challenge to keep pressures in water lines adequate and uniform in a city where altitudes range through 2,000 feet. To do this, the water system is arranged into 16 pressure zones. These are irregular north-south strips away from the river, and each zone is nearly a separate water system. If water is drawn heavily in one area, other water may be pulled from an adjoining zone until pressures at the zone boundaries adjust to predetermined levels. Thus, water in some mains occasionally flows in reverse directions; therefore, water to a given hydrant or faucet may come from different reservoirs or wells at different times.

Wells range considerably in size, depth, yield, and pump capacities, depending on when the well was drilled, the need at a locality, and the yield of water from the aquifer. During 1995 the city wells averaged about 111 million gallons per day, totaling about 40.4 billion gallons for the year. With a service population of about 470,000 people, each person on average consumed about 86,000 gallons in 1996, or approximately 240 gallons a day. In 1996 the basic cost of water was 65.8 cents per 100 cubic feet (748 gallons) plus a small state surcharge, making it some of the cheapest water in the Southwest.

Well fields may have several wells scattered around a reservoir that they feed. A few reservoirs have only one well, others none. Almost all have booster-pump stations, and all have chlorine-treatment equipment. Booster stations at reservoirs contain as many as six huge pumps, some having individual capacities of 7,000,000 gallons a day. Reservoir capacities range from 125,000 to nearly 13,000,000 gallons. The city has a storage capacity of more than 100 million gallons.

NEW MEXICO WATER LAW

The use and appropriation of water is an ancient and fundamental aspect of human existence. Concepts of water rights for kings, individuals, and the public were developed early in history. Pueblo farmers had already engineered extensive irrigation systems and had rules for their use and maintenance, when Juan de Oñate's band of Spanish colonists arrived in New Mexico in 1598. The Spanish came armed with hundreds of years of irrigated-agriculture knowledge and a vast body of land and water law. The Spanish rules of colonization were clear on the interaction between native and colonial irrigation; they stated that the ancient customs and laws of the Indians should be retained and respected as far as practical. Fortunately, many Spanish and Pueblo irrigation practices were similar; for example, the main ditch was publicly owned. In New Mexico the Spaniards adapted their rules to the new land, and ultimately, the acequia system that evolved was a blend of Old World Spanish, New World Spanish, and Pueblo laws and customs. Because of the value of reliable irrigation water, water-user associations became highly integrated and efficient.

Under Mexican rule, community acequia laws remained unchanged. Under U.S. rule, however, new regulations were imposed. Although the Treaty of Guadalupe Hidalgo (in 1848) and the Gadsden Purchase (in 1854) protected water rights of existing acequias, each state or territory now had control of its own water law. In most of the western U.S., British water-law doctrines were modified to incorporate the appropriation and use of water into establishment of water rights. Priority of appropriation and beneficial use

of water according to the law decided the better right. In areas of water scarcity and abundant population, more regulation became necessary to protect the individual and the public and to ensure continued good water supplies for the future.

In 1907 the New Mexico Territorial

Martineztown acequia, ca. 1941. During the 18th century this area was common land where sheep and cattle were grazed by residents of the villa of Albuquerque (now Old Town). Though Martineztown has long since been surrounded by Albuquerque, the neighborhood still retains its identify and its name.

Legislature enacted a water law declaring that all river waters, excluding existing rights, belonged to the public. Today the courts continue to try to sort out the size and validity of rights that existed before 1907. People may use the waters according to legal regulations as long as such use does not infringe upon a prior right. New rights can be granted only upon demonstration by the state engineer that such rights would not be detrimental to existing rights. Land along a river usually carries water rights as a certain number of acre-feet (1 acre-foot = 325,850 gallons) a year for each acre of land owned, provided the water is used beneficially. The state engineer limits the amount of right based on purpose of use, location, nature of the land, known flow of the river, and demonstrated or established need or use of the water. Obviously, there is a limit to the

The riverside drain on the Rio Grande at the west end of Candelaria Road, near the Rio Grande Nature Center.

number of rights that can exist along any particular river.

Prior use of the water determines first right. One does not have a right to use the water merely because it flows by one's land. Land and water rights can be sold independently, but irrigation water rights, if sold, must be transferred to another location for purposes approved by application to the state engineer. Land some distance from a river must also abide by water rights, as pumping from wells commonly affects the water table. In the early days, officials did not recognize the close relationship between water on the surface and that underground; therefore, surface water rights could be bypassed by drilling a well. Thus, a person without legally assigned water rights could drill a well near a river, or even at some distance from it, and get free water that in reality came from the river by seepage.

Fortunately, in 1931 the New Mexico Legislature, upon recognizing the interconnection between surface and ground waters, empowered the state engineer to regulate ground-water usage with river usage. By this "conjunctive use" law, the Office of the State Engineer could (and does) "declare" and outline certain ground-water areas, such as the Roswell artesian basin or the Rio Grande basin, to be under its jurisdiction.

Because of these regulatory powers bestowed upon the state engineer, everyone, including industries, farmers, and many municipalities, must now obtain permission to drill any well in a declared basin. Small domestic wells are exempt from retiring rights. If a city, institution, or industry receives a permit to drill a well, it must retire, or buy and retire, surface-water rights in quantities determined by the size of the proposed well and its proximity to the river, to offset the effects upon the flow of the river. In other words, the well source is exchanged for the river source. If too many wells are drilled next to a river and if these wells deplete the aquifer, the river will begin to lose more water to the aquifer, and the flow of the river will decrease. So, to follow the interstate agreements, and to protect the water rights of surface owners, subsurface-water use must be controlled according to the laws of conjunctive use and prior appropriation.

One might think that this regulation would hamper the development of a city and its industries, but so far this has not been the case. However, as we continue to

This acequia in Placitas channels water from El Oso spring, one of five major springs of Las Acequias of the Community of Placitas. These springs discharge water from the Madera limestone, providing water for domestic use and irrigation. The acequia system here has been in use since the mid-1830s.

increase our demands for water on what is a variable but clearly finite system, some changes are inevitable. These groundwater rights are junior to the much older surface-water rights. And the time could come when their lower priorities mean that pumping must be curtailed whenever the river can't supply the senior rights.

In 1938 New Mexico, Texas, and Colorado reached a formal legal agreement to divide the water of the Rio Grande above Fort Quitman, Texas, 75 miles downstream from El Paso. Division of the water was based upon records of flow along the river in each state; it had the objective of maintaining the status quo of 1929. By the agreement, termed the Rio Grande Water Compact, downstream users are to receive so many acre-feet of water, based on acre-feet of water passing upstream gaging stations. This amount varies from year to year.

Because of over consumption and the vagaries of precipitation, upstream users often owe water to the downstream users. As long as water is owed, it must be released from upstream reservoirs except permanently granted reserves for recreation and fish and bird habitats. The Rio Grande Water Compact is a primary reason for controlling water use in New Mexico.

THE MIDDLE RIO GRANDE CONSERVANCY DISTRICT

The Middle Rio Grande Conservancy District regulates water use (and it contracts with the U.S. Bureau of Reclamation to supply much of the water) for agriculture along the middle Rio Grande.

Although land has been cultivated along the central Rio Grande for thousands of years, the large, modern farms were not possible until the Middle Rio Grande Conservancy District was formed in 1925. Then the conservancy undertook a remarkable engineering plan for an enormous area from Cochiti Lake near Santa Fe to Elephant Butte Reservoir near Truth or Consequences. The plan had three goals: 1) to drain the swamplands along the river; 2) to provide flood control in the

Diversion dam and canal on the Rio Grande at Angostura, looking north.

Rio Grande valley; and 3) to provide irrigation water to farmers. The plan has succeeded spectacularly. Nearly 70,000 acres of bottomland that were swampy in 1920 are now irrigable. However, that success has come at a price. The river channel and bosque ecosystem have been irrevocably changed from a laterally migrating wild river to a channelized and engineered river that no longer floods.

Creation of the district was the result of many years of work by local residents who were determined to solve the problems caused by aggradation of the river bed. Historically, irrigated acreage had increased from about 25,000 acres in 1600 to a maximum of about 124,800 acres in 1880. Subsequently, productive acreage steadily declined to 40,000 in 1925. Valley bottom land was becoming increasingly waterlogged, and spring floods were becoming commonplace. It is thought that the problem began with the 1874 flood, which may have been the largest peak flow seen in historic times on the Rio Grande, estimated at 100,000 cubic feet per second. During the nine days of flooding, the sediment-laden waters destroyed the old river channel and established a new path. Coincidently, irrigation in the San Luis Valley of Colorado greatly increased, resulting in a marked decrease in downstream flow and, therefore, creating a river incapable of moving huge volumes of former floodplain sediment downriver. By 1900 waterlogging of the middle Rio Grande valley had become a serious problem, and several failed attempts were made to organize drainage districts.

The turning point was reached in 1921 when the state created the Rio Grande Survey Commission. The Albuquerque Chamber of Commerce teamed up with the Rio Grande Rotary Club, the Albuquerque Kiwanis Club, and the Albuquerque Board of Realtors to organize a Middle Valley Reclamation Association, designed to cooperate with government agencies in order to secure federal funds for river work. The private association worked with the state commission to pass the Conservancy Act by the state legislature in 1923. Part of the plan required incorporating more than 70 existing acequias into the district master plan. It wasn't until 1956 that the last of these community acequia associations relinquished control of their works to the district.

By 1936 the district had dug 342 miles of drainage canal, 475 miles of irrigation canal, and constructed 200 miles of levees. The major crops in this part of the valley are alfalfa, grain, and other kinds of hay. Permanent grass pastures are also maintained for feeding horses, cattle, and sheep. The farms in the valley could not survive without inexpensive river irrigation water supplied through this elaborate system of supply ditches and drainage ditches. The total value of agricultural production on conservancy-irrigated acreage was more than $32 million in 1999.

Today the middle Rio Grande valley contains some of the most consistently productive farmland in the region, although residential development is expanding and replacing farmland throughout the valley. The conservancy district assesses its members a fee for water delivery. Farmers need about 3 acre-feet per acre per year to grow their crops, but during droughts it may not be possible for the district to deliver that much water. Water users (farmers, municipalities, industries, planners, etc.) and the state engineer have realized for decades that the Rio Grande is over-appropriated, meaning that if everyone demanded all of the water to which they were entitled, there would not be enough water to go around. A group of people representing a great variety of water concerns have recently begun the contentious, arduous, and lengthy process of developing a comprehensive and fair water plan for the middle Rio Grande basin, and there is a call on the legislature and the Interstate Stream Commission to develop a state water plan.

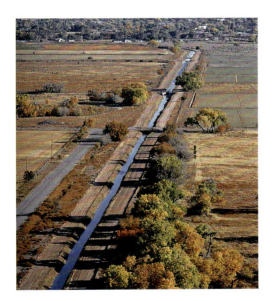

Many issues still need to be faced. The widespread pumping of ground water has prevented use of our basic priority system to determine who must cut back in times of shortage. Pueblo Indian water rights, with a couple of exceptions, are entirely undefined. Domestic wells are widespread, locally use a lot of water, but are subject to no administrative control. Evaporation from Elephant Butte Reservoir and from the river, and transpiration from exotic tree species (i.e., salt cedar and Russian olive) deplete an inordinate proportion of our total water resource. And so far there is no serious reward program for conservation.

The Rio Grande, among all of the rivers in the west, is said to have more bits and pieces of its pre-development environment and biota remaining than any other. Perhaps, if we are careful, in the process of initiating a modernized, equitable water-management system, we can also preserve some of the things that have made New Mexico the Land of Enchantment.

SUGGESTED READING

Ground-Water Resources of the Middle Rio Grande Basin, New Mexico. Circular 1222, U.S. Geological Survey, 2002.

Our Water Resources: An Overview for New Mexicans by William J. Stone. New Mexico Bureau of Mines and Mineral Resources, 2001.

Waters, Rivers and Creeks by Luna B. Leopold, University Science Books, 1997.

Fresh Water by E. C. Pielou, University of Chicago Press, 1998.

The Albuquerque Main Canal flows out of Bernalillo onto the Sandia Indian Reservation. This view is to the north.

CHAPTER FIVE
CULTURAL DEVELOPMENT OF THE ALBUQUERQUE AREA

PALEO-INDIANS

The Albuquerque area contains one of the longest records of human habitation in North America. Stretching back nearly 11,000 years, most of Albuquerque's "history" is prehistory. The earliest known people were big game hunters. These Paleo-Indians traveled over North and South America following migrating herds of giant bison, mastodon, and woolly mammoths. Between kills of these large beasts, they undoubtedly ate plants and smaller game. Evidence of their passing near Albuquerque dates as far back as 8,800 B.C. and includes magnificent stone spear and dart points, called Clovis, Folsom, and Cody points, found at high overlooks and near playas. Over 30 Folsom campsites have been identified near Albuquerque. Very early radiocarbon dates of 17,000–35,000 years ago, derived from Sandia Cave deposits, are now generally believed to represent the remains of animals that used the cave before human habitation, rather than the remains of human activities.

DESERT ARCHAIC

A dramatically different lifestyle, the Desert Archaic, appeared in the Southwest about 5,500 B.C. In some areas of the western U.S., Desert Archaic hunters and gatherers may have been contemporaneous with Paleo-Indian people. Instead of ranging far and wide after big game, Desert Archaic bands of 25–30 people moved around seasonal circuits of perhaps 100–200 miles. They became experts at exploring the food and tool possibilities of nearly every plant and animal in their environment. Desert Archaic equipment reveals this focus—stone dart and spear points for killing game, nets and hunting sticks for capturing birds and small animals, basketry for collection and storage, grinding stones for milling grains and nuts, stone choppers and pounders for preparing roots and other foods, bifacial knives for skinning game and cutting meat and vegetables, and the first slow cookers (cobble-filled roasting pits). Desert Archaic sites are not impressive, usually consisting of a few hearths and roasting pits, a few tools, and the leavings of tool making and food preparation, but their significance is enormous. During the last 1,000 or 2,000 years of the period, Desert Archaic people began tentatively planting domesticated corn and beans from Mexico—a step that would ultimately lead to reliance on domesticated plants and, finally, by about A.D. 400, to a commitment to settled village life. Many Late Archaic (also called Basketmaker II) sites

OPPOSITE:
The mission at Isleta Pueblo.

Large Paleo-Indian point.

from the 800 B.C. to A.D. 400 period have been identified in the Albuquerque area. Although most of them are situated on the West Mesa, in rock shelters, or on gravel terraces above the Rio Grande, at least one site has been found on the floodplain.

> ### AN·A·SA'·ZI
> Anasazi is a term derived from two Navajo words meaning "enemy ancestor" or "ancient enemy." To many of the Pueblo tribes, the word is politically charged and not altogether acceptable. Prehistoric Puebloan and ancestral Puebloan have both been proposed as alternatives. But the term has been in standard use by both scholars and the public for many years. For the sake of clarity (but with a bow to those who might find the term offensive), in the text we use the term Anasazi.

THE ANASAZI

Settlement in villages and towns, of greater or lesser permanence, and reliance on agriculture, to greater or lesser degrees, mark the culture that for years has been known as the Anasazi. The descendants of these people live today in the pueblo villages of New Mexico and Arizona. With settled life and an agricultural base, basket making, one of the hallmarks of the Late Archaic, decreased as pottery making increased. Architecture became more permanent, and ceremonialism became more elaborate. To suggest there was an abrupt switch to dependence on domestic crops and a rush to construct villages when the first Mexican seeds entered the Southwest, however, is an error. The process was long, slow, and tentative. Hunting and gathering require almost constant moving around a seasonal circuit but do not require the massive time investment cultivated fields and permanent houses demand. Nor does the harvesting of wild plants and animals involve the risks of depending on a few garden plots subject to freezing, drought, and insect depredations. Experimentation with raising plants went on for hundreds of years before the final commitment to village life was made. Even then, the commitment was never total. Pueblo people continued to supplement their gardens with wild food until well into the 20th century and can be seen even today harvesting plants for teas and medicines.

BASKETMAKER III–PUEBLO I (A.D. 400/450–900) In the Four Corners area, the break between these two periods is

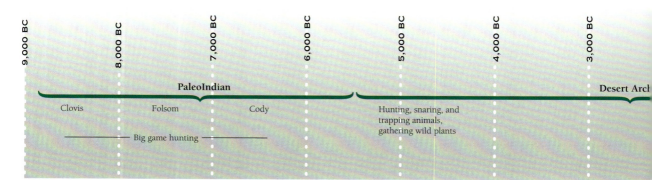

CULTURAL DEVELOPMENT

Early Basketmaker seed pouch.

sharply defined, but sites along the Rio Grande often contain elements characteristic of both periods. This phase is characterized by pit houses clustered in groups of two to 20 or more. Surface storage structures began to appear later in the period, but pit houses remained the domiciles. Agriculture based on corn, beans, and squash provided perhaps half the food supply. The excellent basketry of the Late Archaic (Basketmaker II) continued for a time but soon gave way to ceramics—graywares and black-on-white painted wares. The bow and arrow arrived from the north about A.D. 400, but chipped stone working became more casual, even as seed-grinding tools became more formal.

Known sites of this period are located on gravel terraces near Sandia Pueblo, on a hilltop in Tijeras Canyon, and on low terraces and sandy hills west of the Rio Grande. Proximity to water, either the Rio Grande or its tributaries, is the unvarying topographic context.

PUEBLO II (A.D. 900–1200) This was the period of the dramatic building boom in Chaco Canyon and elsewhere throughout the San Juan Basin. Engineered roads, massive multistory masonry pueblos, elaborate ceremonial structures (the great kivas), trade with Mexico in such exotic goods as parrots, and an immense number of new black-on-white ceramic design styles characterized Pueblo II in the San Juan Basin.

In the Albuquerque area, however, life was essentially unchanged from that of Basketmaker III–Pueblo I times. People continued to live in pit houses and store their food, both cultivated and wild, in surface storage rooms. Except for minor additions, artifacts remained the same. Only a few Pueblo II sites are known for the Albuquerque area. Their topographic situations are the same as those of Basketmaker III–Pueblo I sites.

PUEBLO III (A.D. 1200–1300) Like Chaco Canyon and the San Juan Basin during Pueblo II times, Mesa Verde (in Colorado)

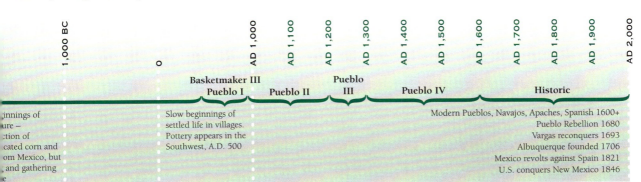

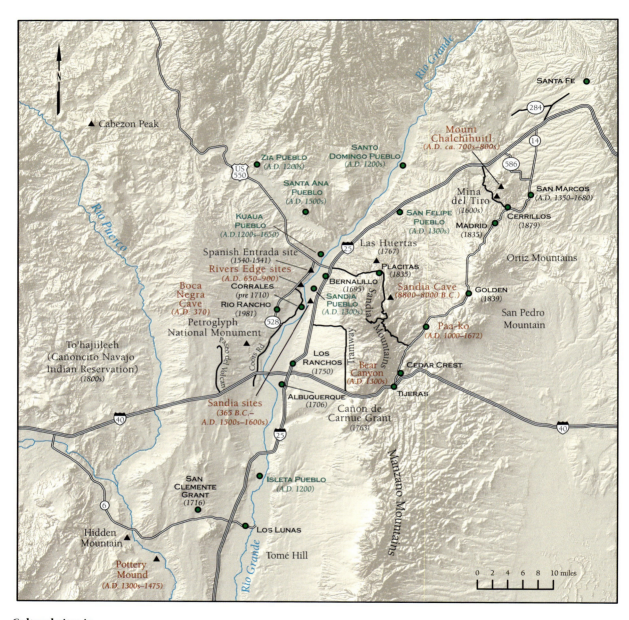

Cultural sites in the Albuquerque area. Prehistoric sites are shown in red, pueblos are shown in green.

was the star of Pueblo III and repeated similar architectural, religious, ceramic, and demographic patterns. Nothing so spectacular occurred in the Albuquerque area, although important changes did take place. The most significant is that the number of habitation sites increased, many of them immediately adjacent both to major drainages and to arable land. Architecture was varied, both pit houses and surface roomblocks functioning as habitations. Rectangular kivas built into roomblocks served ceremonial needs. Also of real importance is the sudden abundance of varied ceramic types, evidence of distant trade relationships. Besides locally produced pottery, black-on-white wares were imported from the Santa Fe and Galisteo Basin areas to the north and the Socorro area to the south. From the west came small amounts of polychrome wares.

Pueblo III sites in the Albuquerque area are more numerous than Pueblo II sites and are located on the Rio Grande floodplain, in Tijeras Canyon, and near Taylor Ranch and Corrales. Isleta Pueblo, still a thriving village south of Albuquerque, was founded during Pueblo III. Present Isleta Pueblo was settled in the late 1600s or early 1700s, but several nearby ancestral village sites are dated (via ceramics) to A.D. 1200.

PUEBLO IV (A.D. 1300–1600) By Pueblo IV times, both Chaco and Mesa Verde had been abandoned, and the population had moved to permanent water sources—major streams or, at least, seeps and springs. A long stretch of the Rio Grande, from Taos to Socorro, was inhabited. Many residents lived in large towns of several hundred rooms next to the Rio Grande, some in villages of 50 or more rooms, and others in tiny one- to four-room hamlets. Pueblo IV sites were frequently abandoned or washed out.

Pueblo IV has been called the "Rio Grande Classic," a period characterized as one of "general florescence of material culture..., expressed by features such as the elaborate decoration of pipes, elaborate axes, numerous vessel forms, carved bone tools, stone effigies, and mural paintings." Use of a new substance for decorating pottery (lead glaze) appeared, probably from the Zuni area, but most of the early Rio Grande glazewares were locally produced. The appearance of the new Rio Grande style, with its emphasis on masks in rock art and kiva murals, may reflect introduction of the kachina complex from the Jornada Mogollon area (Las Cruces northeast to Corona). Hundreds of representations of masks were carved into the basalt faces of Albuquerque's West Mesa escarpment, now Petroglyph National Monument, during Pueblo IV times. Portions of the magnificent murals from the main kiva at Kuaua [KWAH-wah] are on exhibit at Coronado State Monument. They show masked ceremonial figures, animals, and other elements important to Pueblo religion

Awanyu bowl from Pecos Pueblo, ca. 1600-1700.

of the 1400s and 1500s.

When the Spanish arrived in the Albuquerque area, they found a dozen or more villages inhabited by Tiwa speakers (the language still spoken at Sandia, Isleta, Picuris, and Taos) and named it the "Tiguex Province." Sandia Pueblo, just north of Albuquerque, was one of the villages. It was founded during the Pueblo IV period and, except for a brief period of abandonment after the Pueblo Revolt, has been occupied almost continuously since the 1300s. At least 30 abandoned Pueblo IV sites exist near present Albuquerque, on the floodplain east and west of the river, on bluffs above the river, and in Tijeras Canyon.

The last remaining large Pueblo IV site (except Kuaua) along the Rio Grande near Albuquerque, Piedras Marcadas Pueblo, is now part of Petroglyph National Monument. Farther afield is Pottery Mound, approximately 30 miles southwest of Albuquerque. This large pueblo consists of three superimposed villages, portions of which once stood three to four stories high. Renowned for its fabulous kiva murals, Pottery Mound has not been reconstructed and cannot be visited, but copies of some murals have been reproduced in print (see Suggested Reading at the end of this chapter).

17th century Spanish spur, found at Pecos National Historical Park, New Mexico.

ARRIVAL OF THE DINÉ & SPANISH

Two new groups arrived in New Mexico in the 1500s, one from the north and the other from the south. The first group, the Diné, whose descendants are known to us as the Navajos and Apaches, had been hunting and gathering their way south from Canada for many years before they reached New Mexico—and presumably stopped then only because the presence of settled people seemed a barrier to those who must rely on uncontested land for hunting and collecting plant foods. They probably traveled in small bands and may have reached the upper San Juan Basin in the mid-1500s, or perhaps as early as the 1400s.

The second group came north in 1540 from Mexico City, led by Francisco Vásquez de Coronado in search of the Seven Cities of Cibola. When Coronado found the Zuni villages to be constructed of stone rather than gold, he moved to the Tiguex Province on the Rio Grande, where he spent the next two winters. Until recently it was believed that the Pueblo of Alcanfor, in and near which the Spaniards camped, was Kuaua (at Coronado State Monument), but excavations conducted in 1986 show a Spanish encampment nearly 2 miles to the south near a pueblo now called Santiago. Metal artifacts and sheep bone found mixed with other plant and animal remains, including pueblo ceramics dating between 1525 and 1625, strongly suggest that this was part of Coronado's expeditionary encampment. If so, it is one of the earliest Spanish sites known in North America. Alcanfor also figured in the first Spanish/Indian tragedy in New

CAÑONCITO NAVAJO RESERVATION

The exact time of the arrival of Diné people in the Southwest is not known, but it was apparently not long before Coronado's entrada in 1540. These Athapaskan speakers, who later became the Navajo and Apache, find their closest linguistic relatives in Alaska, Canada, and on the Northwest Coast. Spanish documents record Navajos in the Cebolleta–Encinal area (northwest of present Laguna Pueblo) in the 1700s, and perhaps as early as the 1580s.

In 1748 the Catholic Church built missions at Cebolleta and Encinal and invited several bands of Navajos to move back from the north to live near the missions. Perhaps 500 people accepted, but by 1750 the missions had begun to fail, and many Navajos moved away. Over the next century the Cebolleta Navajos (also known as "Sandoval's band") continued to plant corn and graze sheep and horses throughout the Cebolleta Mesa–Rio Puerco area, moving their herds as grass and water supplies required. When food supplies became scarce, friendly relations with the pueblos of Laguna and Acoma and nearby Hispanic villages changed to raids and attacks.

Civil War battles in New Mexico distracted the attention of the U.S. military, and Navajo and Apache raiding intensified to the point that in 1862 Governor Henry Connelly and Brigadier General Carleton placed Colonel Kit Carson at the head of a U.S. military unit, charging Carson with incarcerating all Navajos and the Mescalero Apaches. Carson obeyed reluctantly, believing that the Indians would agree to a treaty with the U.S. without war. Subduing the Mescaleros after continual battles in 1862 and 1863, Carson turned his attention to the Navajo, pursuing them throughout their vast New Mexico–Arizona homeland. Finally, in January 1864, he breached the one place Navajos considered safe and impregnable, Cañon de Chelly in northwestern Arizona. He set peach orchards and stored crops ablaze, confiscated sheep and horse herds, and instructed the people to gather at Fort Wingate near Gallup to begin the "Long Walk" to Fort Sumner (Bosque Redondo) in eastern New Mexico.

More than 9,000 Navajos were held at Fort Sumner, a place of swampy infertile soil and little wild game or plant foods. The Mescaleros escaped, and untold numbers of Navajos died of disease and malnutrition. The U.S. declared Bosque Redondo a disaster and allowed the Navajos to return home in 1868. When they reached the Cebolleta area, small groups (perhaps a total of 400

people) began leaving the main body, some to rejoin family members who had been succored by Laguna Pueblo friends. Between 1869 and 1881, a few families began moving east to the Cañoncito area, and in 1886 a group of 19 Navajo settlers filed claims under the Indian Homestead Act of 1884. More families moved to the area between 1912 and 1914, and the Cañoncito Navajo Reservation was ultimately formed.

Mexico. When the New Mexico winter proved harsh, Coronado moved the residents out of the pueblo and his troops in. The people of Alcanfor scattered to nearby villages, leaving their storerooms full of food and supplies to the Spaniards. Not only did the troops consume Alcanfor's supplies and call on other pueblos' stores, but they also attempted to seduce Pueblo women. That seemed the final insult, and a party of Indians drove off the army's horse herd and killed a guard. Retaliation was instantaneous. The Spaniards attacked several Tiwa villages, looting supplies, torching houses, and murdering inhabitants, even burning some at the stake. Outrages against other Pueblo villages followed. It was not a good beginning. Finally, after searching from New Mexico to Kansas and finding no gold, Coronado returned to Mexico in April of 1542, and New Mexico was all but forgotten for nearly 60 years.

A few forays were made into New Mexico in the 1580s and '90s, but it was not until 1598 that Don Juan de Oñate was granted permission to bring colonists and form permanent settlements. The official rationale was that New Mexico was filled with souls that needed saving, but protecting the rich silver mines in northern Mexico may have added a sense of urgency to the religious impulse. Spanish settlements in New Mexico would provide a buffer against French adventurers from the north.

By the time Spanish settlement began in New Mexico, Spain had gained extensive experience in New World colonizing. Slave taking (unless Spaniards had been directly attacked) had been outlawed in 1542. Native water rights received precedence and must be respected; Spanish citizens were forbidden to encroach on Pueblo lands in any way; a special court to which Indians had direct access (Juzgado de Indios) was given jurisdiction in civil and criminal proceedings, and a special official, Protector de Indios, was appointed to aid and advise them; Indians were exempt from the Inquisition.

Spanish and Indian relations seem to have been amicable initially, but harmony was short-lived. The institutions of encomienda and repartimiento, both probably lifted directly from feudal Europe, were installed. An encomendero, always a prominent and trustworthy citizen, was granted the right to collect taxes from a pueblo in return for providing education, aid, and protection from enemy attacks. The system of repartimiento allowed a Spanish landowner to hire, for wages and daily rations, a group of Indians conscripted by the governor for brief periods. As for the clergy, their twin aims were conversion to Catholicism and teaching such European skills as stock raising and carpentry. In return, they expected donations of both goods and labor for maintaining the missions—something the members of the pueblos had long been accustomed to providing to their own religious organizations.

Despite good intentions, however, cultural misunderstandings and unforeseen events combined to create disaster. Instead of benign and gentle guidance and infrequent wage labor, the encomienda and repartimiento systems were sometimes abused, becoming a drain on the pueblos' labor, food, and manufactures. Although both civil and religious officials were theoretically pledged to protect Indians, individual personalities sometimes clashed, trapping Indian groups in the middle. Unfortunately, early governors were selected more for military skills than for administrative and diplomatic abilities. Catholicism, a state religion based on a contract between a single individual and a single god, proved incompatible with Pueblo religions, which were based on a contract between an entire group and a large pantheon of gods.

Plains tribes had long brought such items as dried buffalo meat, tanned robes, and river shells to the big trading pueblos of Taos, Pecos, Gran Quivira, and Quarai to exchange for corn, salt, pottery, and cotton textiles. Fearing the influence of the free tribes on the pueblos, the government periodically forbade trading with them. Spanish and Pueblo governmental institutions fit no better than any of the other institutions. Each pueblo was essentially a city-state, independent of all other pueblos. They were not accustomed to rule by a central authority. Finally, two accidents of nature added to the problems. An epidemic of European diseases—smallpox, whooping cough, and measles—reduced native populations quickly and tragically. A lengthy drought in the 1670s produced serious famine in Pueblo villages, changing sporadic raiding to almost perpetual raiding by equally hungry Apaches and other Plains tribes.

PUEBLO REBELLION

In August 1680 pueblos from the Rio Grande to the Hopi mesas in Arizona banded with Apaches and perhaps Navajos to burn churches, archives, and houses and murder Spanish priests and citizens. The Spanish settlers fled down the river to El Paso, where they remained for more than a decade. Don Diego de Vargas Zapata y Luján led 800 settlers back to Santa Fe in 1693, but the Spaniards had learned a major lesson from the shock of the Pueblo Rebellion. The practice of encomienda was ended permanently. Catholic priests learned to accept the many values in Pueblo religions and encouraged the continuation of old practices side by side with Catholicism, as is true today. Governors were selected for expertise in humane administration, rather than in warfare.

Spanish retablo ca. 1795 – 1820, Pecos National Historical Park, New Mexico.

LANGUAGES

Languages from several different language families, besides Spanish and English, are spoken in the Southwest. One major language group is Tanoan, which is broken into three sub-groups—Tewa, Tiwa, and Towa. Tanoan languages are spoken at 11 pueblos in New Mexico and one pueblo in Arizona (Hano, on the Hopi mesas). Sandia and Isleta Pueblos, on the north and south borders of Albuquerque, both speak Tiwa. Keresan, another major group, is spoken at seven pueblos in New Mexico, including Laguna and Acoma west of Albuquerque. Zuni, in extreme western New Mexico, is a language isolate unrelated to any other known language, with the possible exception of Penutian, spoken on the Northwest Coast. The remaining Pueblo language is Hopi, spoken in the Hopi villages of northern Arizona. Hopi is a Uto–Aztecan language and is related to the languages of the Utes, Southern Paiutes, and other groups in the western U.S. and to Nahuatl, the language of the Aztecs in the Valley of Mexico. Unrelated to any of the others is Southern Athapaskan, spoken by the Navajos and Apaches, whose linguistic relatives are Northern Athapaskan speakers in Canada, Alaska, and the Northwest Coast.

Language Family	Language Group	Language/Speakers
Athapaskan	Southern Athapaskan	Navajo, Apache
Kiowa-Tanoan	Tiwa	Taos, Picuris, Sandia, Isleta
Kiowa-Tanoan	Tewa	San Juan, Santa Clara, San Ildefonso, Pojoaque, Nambe, Tesuque
Kiowa-Tanoan	Towa	Jemez
Keresan	Keres	Cochiti, Santo Domingo, San Felipe, Zia, Santa Ana, Laguna, Acoma
Zuni	Zuni	Zuni
Indo-European	Romance	Spanish
Indo-European	Germanic	English

Most pueblos welcomed the return of the Spaniards, for the Utes and Comanches now had large horse herds and engaged in repeated raids on pueblo villages. Protection by the Spanish military offered much-needed relief.

People coming to New Mexico for the first time are often shocked to learn that Apaches, and probably Navajos, joined the pueblos in rebelling against Spanish rule and, even more surprising, that Pueblo people who feared retaliation for their part in the rebellion fled to the Navajos and Apaches for safety when the Spaniards returned. The image of Navajos and Apaches in the popular press has been one of ruthless and unceasing raiders of hapless pueblos. A picture closer to the truth is revealed by both Spanish documents and archaeology. The mutual benefits of trade had produced relationships that seem to hold true the world over between mounted tribes and sedentary villagers. When times were good, a Navajo or Apache band might camp for weeks, or even months, directly outside a pueblo—trading, joining hunting parties, attending the pueblo's feast days, exchanging gifts, and probably forming friendships. When times were bad, the pueblo might refuse to trade, fearing their own food shortages. If a Navajo or Apache band became desperate enough, they reverted to raiding, since they knew where the dried corn, squash, and beans were stored. The following year might find them back camping on a pueblo's doorstep. Given such a relationship, clearly the Pueblo people who turned to the Navajos and Apaches for succor were not flying to enemies who would chop off their heads without warning.

In the 1690s Spanish settlers began to establish homes and farms on both sides of the Rio Grande near Albuquerque in areas either abandoned or never claimed by pueblo groups. The Hacienda de Mejía, south of the later site of Albuquerque, was a designated paraje, an overnight stopping place for travelers on the Camino Real, the royal road that stretched from Santa Fe to Mexico City. Albuquerque was founded by Governor Francisco Cuervo y Valdés in 1706. Construction of the church, San Felipe de Neri, began on the west side of the plaza (the present church, on the north side of the plaza, was started in 1793). The plaza at Albuquerque soon became an important gathering point. Alcaldes (mayors) throughout New Mexico were notified in 1732 that merchants bound for the trade fairs in Durango and Chihuahua were to rendezvous at the plaza. The troops employed by Kit Carson in the Navajo roundup, which began in 1863, were mustered out in the Albuquerque plaza in 1865, and the Navajos returning from Bosque Redondo (Ft. Sumner) in 1868 camped just west of the plaza while they waited to be ferried across the river.

MEXICAN & U.S. RULE

In 1821 Mexico revolted against Spanish rule, and New Mexico was opened to traders from the U.S., who poured goods into the new and enthusiastic market. The resources of the new government in Mexico City proved to be stretched so

UNIVERSITY STYLE

The Pueblo style so apparent at UNM today does not necessarily reflect the evolution of architecture on campus. Hodgin Hall, the first building at UNM, was completed in 1892, a three-story red-brick Victorian structure. In 1908 UNM president Dr. William Tight remodeled the structure, in accordance with his policy that all university buildings should be Pueblo style. An enraged Board of Regents fired him in 1909. Most of the original building is still present under the stucco.

Hodgin Hall, 1890 (left); after remodelling, 1908 (below).

thinly over a vast area that when the U.S. invaded New Mexico in 1846, New Mexico quickly fell to U.S. forces. From then on, major events touching Albuquerque were tied to the fortunes of the U.S.

New Mexico's share of the Civil War included Confederate General H. H. Sibley's 2-month occupation of Albuquerque in 1862, from whence he hastily departed to fight (and lose) the definitive battle at Glorieta Pass near Pecos. One of the Civil War cannons he buried in Albuquerque during his retreat is now on display at the Albuquerque Museum in Old Town. The Atchison, Topeka, and Santa Fe Railroad (AT&SF) reached Albuquerque on April 5, 1880, and "New Town," east of the Old Town plaza, was soon constructed. Saloons, gambling halls, brothels, and even opium dens were rapidly established in the midst of the quiet community. It took more time to establish a university. A bill creating the University of New Mexico was passed in 1889, but the first building—isolated in the Northeast Heights at a distance from the town—was not completed until 1892.

THE RAILROAD & ROUTE 66

Both the coming of the railroad in 1880 and the construction of Route 66 caused dramatic changes in Albuquerque. The fervent promotion of mass tourism by the AT&SF—aided and abetted by Fred Harvey and his hotels, dining rooms, and Harvey Indian Detours—also played an

ROUTE 66

U.S. Highway 66 was proposed in 1926. The original route followed the railroads and the Camino Real. From Santa Rosa, the route left the Chicago, Rock Island and Pacific Railroad and paralleled the Pecos River to Romeroville. From there it followed the Santa Fe Trail and the Atchison, Topeka and Santa Fe Railroad into Santa Fe, where it turned south, made its tortuous way down over the basalt escarpment of La Bajada Hill, and crossed Tonque Arroyo just east of present I–25. The "Big Cut" on San Felipe Pueblo land south of the arroyo (visible today from I–25) was constructed before 1926, an engineering feat of the day—60 feet deep and wide enough for a bus. Reconnecting with the old El Camino Real route, the road passed through Algodones, Bernalillo, and Albuquerque. In Albuquerque, it

crossed Barelas Bridge to the west side of the Rio Grande and continued south down NM–1 (now Isleta Boulevard) through Isleta Pueblo to Los Lunas. It then turned west, following the AT&SF Railroad along NM–6 to Laguna Pueblo and continued onto Gallup and into Arizona.

The Laguna Cutoff was built as a state road in 1926. With the construction of Old Town Bridge across the Rio Grande in 1931 and the Rio Puerco Bridge in 1933, more traffic chose that route. However, it was not until 1937 that the federal government adopted the Laguna Cutoff, incorporating it into U.S. Highway 66 when the route was straightened to run west from Santa Rosa across the Estancia Valley to Tijeras and Albuquerque.

important role. When Route 66 was established in 1926, the route paralleled the railroad tracks and the Camino Real (old U.S. 85, now NM–47/2nd Street) through Albuquerque; early motorists relied on railroad-related businesses along the 2nd and 4th Street commercial districts. After 1937 the original route (north from Santa Rosa to Romeroville near Las Vegas, west to Santa Fe, and south to Albuquerque) was abandoned, and Route 66 ran due west from Santa Rosa to Albuquerque. Auto campgrounds, tourist courts, gas stations, cafes, stores, and curio shops materialized almost overnight along both east and west Central Avenue. Residential development followed. Albuquerque was soon transformed from a linear to a cruciform town, unconsciously echoing a pattern that began 11,000 or 12,000 years ago when Paleo-Indian hunters and later hunter/gatherers followed the natural north–south and east–west routes. Later still, ancestral Pueblo people and Spanish settlers planted their gardens and built their villages near the streams and springs that flowed along these routes. On August 10, 1999, Congress authorized $10 million for restoration of the mother road. Albuquerque is fortunate in having 19

Downtown Albuquerque, 1881.

The Santa Fe Railroad depot and the Alvarado Hotel, ca. 1930.

View to the east along Central Avenue, downtown Albuquerque, 1931. The Kimo Theatre, built in 1927 in Pueblo Revival Art Deco style, has been restored and still stands at 423 Central Avenue NW.

Route 66-related properties listed on the National Register of Historic Places.

ALBUQUERQUE TODAY

Albuquerque remained a small town until after World War II. Development of the atomic bomb at Los Alamos led to the establishment of research and support facilities at Kirtland Air Force Base, Sandia Base, and Sandia Laboratories on the southeast edge of the city. Albuquerque had been famous for its healthful climate and medical facilities (especially for victims of tuberculosis) since before the turn of the century, and the health industry again burgeoned. Education, scientific research, and health facilities continue to employ major segments of a growing population.

Albuquerque's diverse citizenry has long been a source of pride. Besides the pueblos of Isleta on the south and Sandia on the north, at essentially the same locations as they were in 1540, Albuquerque contains many descendants of the original Hispanic settlers. Railroad construction and shops brought Anglos, Italians, Blacks, and Chinese in the 1880s. More recently, the city has welcomed newcomers from the eastern and western U.S. and Indochina.

Each era of the city's past can be traced in its architecture. Besides Old Town, parts of the South Valley and North Valley retain houses and churches of the old Spanish villages now engulfed by Albuquerque—Barelas, Santa Barbara, Martineztown, Los Duranes, Los Griegos, and others. Huning Highlands reflects the "New Town" of the 1880s, an area of Victorian gabled roofs and gingerbread trim. Shaded streets and occasional brick houses in parts of the university area recall the eastern campuses from which the original faculty came, although the university itself, until recently, adhered to Spanish–Pueblo style architecture. The Kimo Theater, on Central Avenue, built in 1927 and rescued from the demolition ball in the late 1970s, has been restored to its original splendor as a Pueblo Revival Art Deco masterpiece. Newer sections of the

ALBUQUERQUE INTERNATIONAL BALLOON FIESTA

In 1972 thirteen bold balloonists gathered in the Coronado Center parking lot at the first balloon fiesta. The following year, 13 countries (138 balloons) competed in the "First World Hot Air Ballooning Championships" at the state fairgrounds. By 1978 the 273 balloons represented the largest ballooning event in the world. By 1998 the fiesta had expanded to 873 balloons from 16 countries and 41 states, including 83 special shapes. Event statisticians reported that during the week-long spectacle, balloonists made 6,637 flights that were viewed by 1,067,036 spectators.

city, on both the east and west mesas, conform to the generic fashions of each particular construction phase.

As new industries and new people arrive, Albuquerque continues to grow. Having expanded north to Sandia Pueblo, south to Isleta Pueblo, and east to the Sandia Mountains, development is moving west. Already over the volcanic escarpment and past the volcanoes, it will presumably continue west through the vast Atrisco and Alameda Grants until it reaches the Rio Puerco, to be blocked finally by Cañoncito Navajo and Laguna Pueblo lands.

SUGGESTED READING

Albuquerque: A Narrative History by Marc Simmons. University of New Mexico Press, 1982.

Ancient Peoples of the Southwest by Stephen Plog. Thames and Hudson, 1997.

Behind Painted Walls: Incidents in Southwestern Archaeology by Florence Cline Lister. University of New Mexico Press, 2000.

The Coronado Expedition to Tierra Nueva by Richard Flint and Shirley Cushing Flint. University Press of Colorado, 1997.

Hispanic Albuquerque by Marc Simmons. University of New Mexico Press, 2003.

Rio Del Norte: People of the Upper Rio Grande from Earliest Times to the Pueblo Revolt by Carroll J. Riley. University of Utah Press, 1996.

Great River: The Rio Grande in North American History by Paul Horgan, Texas Monthly Press, 1984.

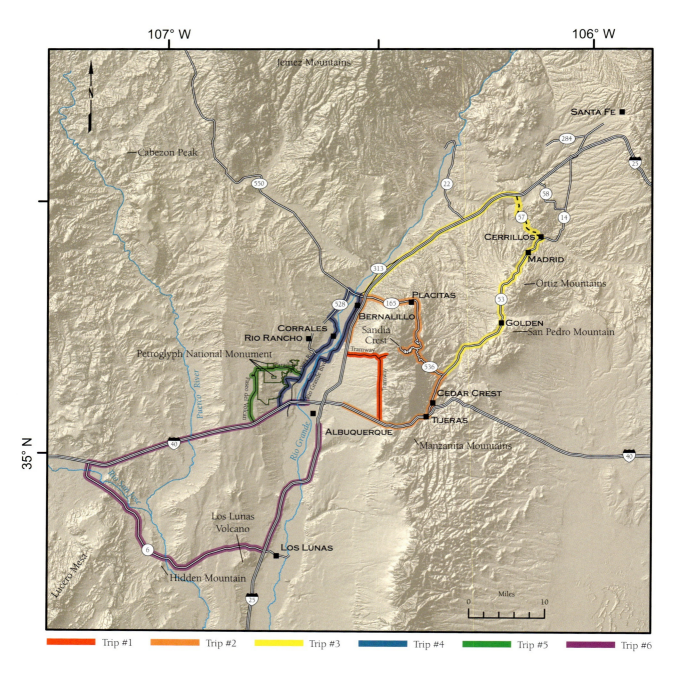

CHAPTER SIX
THE SCENIC TRIPS

The trip logs included within this book are loop excursions that begin and end in Albuquerque. The logs are designed to be used with your vehicle's trip meter: cumulative mileage is shown in the black bold type at the beginning of each entry; the mileage between points is shown at the end of each entry. Because odometer readings vary from vehicle to vehicle, check-point mileages such as mile markers, cattle guards, and intersections are included and should also be used as a guide. Many of the six trips in this chapter have been broken into smaller legs, in order to allow more flexibility for the user.

Critical directions, warnings, and instructions are printed in bold, green type. Features of interest are pointed out using the clock system. For example, the front of the car is 12:00 (we generally say "straight ahead"), due left is at 9:00, and due right is at 3:00. If an object is identified at 1:00, it is just to the right of straight ahead.

Many of these trips are on public land. Please remember that, in most cases, collecting of any sort is prohibited on public lands. Be particularly careful around archaeological sites: they are protected by law, and (more importantly) they are fragile, and scarce. In places where the road crosses private land, please respect landowners rights, and obey all posted signs.

For obvious reasons of driving safety, these trips are designed for at least two people: it is the passengers rather than the driver who must follow the odometer and read the logs. Try to read far enough ahead of your location so that the description can be completed before you reach the point of interest. Be sure that you have water and a full tank of gas before starting out. Gas stations are scarce along some of the routes. When stopping to look at rocks, beware of rattlesnakes and scorpions. Always turn a rock before picking it up. Never stop in the middle of a road, and pull off the road only where time, space, or designated pullouts allow. Be careful of traffic, cyclists, and pedestrians when pulling off the road. Enjoy the trips, and—above all—drive carefully!

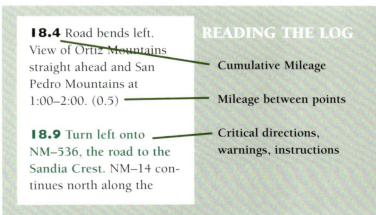

READING THE LOG

18.4 Road bends left. View of Ortiz Mountains straight ahead and San Pedro Mountains at 1:00–2:00. (0.5)

— Cumulative Mileage
— Mileage between points

18.9 Turn left onto NM–536, the road to the Sandia Crest. NM–14 continues north along the

— Critical directions, warnings, instructions

SCENIC TRIP ONE - (20.3 MILES)
The Western Flank of the Sandia Mountains

This tour examines the piedmont deposits and Precambrian rocks exposed along the western front of the Sandia Mountains. Included is a log describing features observed while riding the Sandia Peak Tramway to Sandia Crest. Along the tour route, flash flooding has been a major hazard, claiming lives and destroying property. Methods used to safeguard against floods will also be discussed.

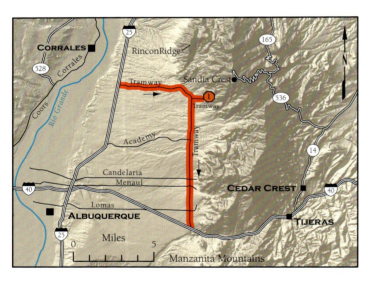

The tour begins at the intersection of I–25 and Tramway Boulevard and heads up the alluvial apron on Tramway Boulevard to the Sandia Peak Tramway. The tram offers visitors a thrilling ride to the top of the Sandia Mountains year-round, for a fee. Our trip continues along Tramway Boulevard south across several flood-prone arroyo channels to the I–40 interchange. Optional stops are included at Juan Tabo Recreation Area and at Albert G. Simms City Park in Bear Canyon. The alluvial-fan deposits and the Precambrian rocks can be examined at the two optional stops and near the tramway. Driving time for this tour is about 2.5 hours without stops. Allow at least an hour for the tram ride.

Exit I–25 at Tramway Blvd (NM–556) and drive east toward the mountains. Log begins at milepost 2.

0.0 MP 2. For the next 5 miles we ascend the Llano de Sandia. Between here and the Sandia Peak Tramway, the land gains about 1,000 feet of elevation. The Llano de Sandia is composed of alluvial fans that have merged to form a continuous westward-sloping plain along the Sandia Mountain front. The surface has been dissected by numerous subparallel, westward-flowing arroyos. Pleistocene alluvium underlies the ridges, and younger Holocene alluvium fills the active arroyos. Vegetation on the surface is dominated by shrubs and grasses with scattered prickly pear and cholla cacti. Be on the lookout for some of the local residents: Jack and cottontail rabbits, coyotes, turkey vultures, and hawks. (0.4)

0.4 Traffic light at Rainbow; Sandia Casino on the left, Bien Mur Indian Market Place on right. Note the American bison in the fenced area. We are on the Sandia Pueblo Indian Reservation; otherwise this region would already have become a suburb of Albuquerque. This is how eastern Albuquerque looked before urbanization. (0.6)

1.0 MP 3. With good lighting, the twin steel cables on which the tramway travels are visible straight ahead. The Tiwas call the Sandia complex Bien Mur, "big mountain"; the Tewas call it Oku Pin, "turtle mountain." The Navajo name for the Sandia Mountains means "revolving (in a horizontal plane) mountains." The Sandia Mountains figure in the mythology of all the Indian groups. (1.0)

2.0 MP 4. Rincon Ridge at 10:00–11:30 is composed of Precambrian metamorphic rocks (schist and quartzite) intruded by fine- and coarse-grained dikes. This sequence is also intruded by the Sandia granite.

From geophysical studies we know

SANDIA PUEBLO

When the Coronado expedition first saw Sandia Pueblo in 1540, chroniclers did not record its name or exact location, but pottery on trash mounds near the modern village indicates that the pueblo must have been near its present location by the early 1300s. Sandia Pueblo joined in the Pueblo Revolt of 1680. After fending off attackers in Santa Fe for 10 days, Governor Antonio de Otermín led the straggling survivors south from Santa Fe. Twenty-one priests and 380 civilians had been murdered. He reached Sandia and found the arms hacked off the statue of St. Francis, the priests' vestments and holy vessels desecrated, and the church on fire, but he continued on. Just south of the pueblo, a group of armed Tiwas attacked the caravan, but when Otermín returned fire, they vanished into the hills, driving captured Spanish horse and cattle herds before them. Turning for a last look over his shoulder and seeing the column of smoke rising from the church, Otermín lost his composure. He ordered the entire village burned—and burned it again when he attempted a reconquest in 1681. The village was empty when reconquest attempts were made in 1688 and 1689, and the houses were in ruins when the Spaniards returned for good in 1692.

Where the Sandias went is uncertain. They may have scattered to other pueblos, perhaps as far away as the Hopi villages in Arizona. Sandia was not reestablished until 1748, when refugees from several pueblos moved in. By 1760 a group from Hopi had joined the village. The Hopis were ultimately absorbed, but Hopi ceremonies are observed at Sandia today. Census figures reflect these events at Sandia, for the population dropped from 3,000 in 1680 to 350 in 1748. After a long period of losing population, the Sandia figures are rising again (471 in December 1998). A few Sandias today practice farming, herding, and jewelry making, but most are employed outside the pueblo in white-collar jobs or in skilled professions.

Sandia Pueblo, 1880.

The name Sandia (Spanish for watermelon) was inspired by the rosy appearance of the nearby mountains at sunset. The Sandias' own name for the village is Naphi'ad (at the dusty place), the origin of which becomes all too apparent when a spring windstorm is howling.

that a very large, buried fault in this area separates the Sandia block from the Albuquerque Basin. An oil-test well, the Norins Realty North Albuquerque Acres No. 2, supports this interpretation. The well was drilled about 2.5 miles southwest of here in 1940 to a total depth of 5,024 feet. The hole never hit bedrock, indicating that the basin-fill deposits are at least 5,000 feet thick within a few miles of the mountain front.

3.0 MP 5. Television and radio towers visible on Sandia Crest at 11:00. The large building at the base of the mountain houses the terminal station for the tramway. Along the skyline, the second tower of the tramway cable system is visible. (1.0)

4.0 MP 6. Junction with FR 333; entrance to Juan Tabo Recreation Area, established by the U.S. Forest Service in 1936. FR 333 leads to picnic areas, restrooms, and

THE SANDIA GRANITE

The Sandia granite is part of a vast complex of granitic rocks exposed in western North America that range in age from Precambrian to Cretaceous. It is one of the most highly studied rocks in the Southwest. Dozens of samples of Sandia granite have been collected for geochemical analyses and isotopically dated in attempts to characterize its age and origin. It appears that the granite crystallized about 1.43 billion years ago. Technically, the Sandia granite is not a granite at all. True granites have 20 to 60 percent quartz and a ratio of alkali feldspar/total feldspar between 65 and 90 percent. The Sandia granite is quite variable in mineralogy, but on average consists of 35 percent quartz, 15 percent microcline, 35 percent plagioclase, 10 percent biotite, and 5 percent microperthite. This makes it a quartz monzonite to granodiorite. However, true granite compositions are rare throughout the world, so geologists tend to use the term granite broadly for any crystalline, quartz-bearing plutonic rock.

Note the numerous "jumping cholla" (Opuntia fulgida) along both sides of the road. These common desert cacti can grow up to 15 feet tall, with a trunk diameter of 10 inches. In early summer large red to purple to pink flowers adorn the cholla. Yellow fruits remain on the plant through the winter. The jumping cholla does not really jump, but its barbed segments are easily detached from the plant and impale the skin with even the most gentle contact. (1.0)

hiking trails within the Cibola National Forest. The La Luz Trail begins in the picnic grounds and ascends the rugged face of the mountain to Sandia Crest. Each year competitive runners on this trail make a grueling 8-mile climb to the summit.

The identity of Juan Tabo is a mystery, even though numerous features around Albuquerque bear his name. One legend holds that Juan Tabo was an Indian sheep herder who grazed sheep near here. Another states that he was a priest who lived nearby. We'll probably never know the truth.

The round-trip mileage to the picnic area is 4.8 miles. Precambrian metamorphic rocks (mostly muscovite-rich gneiss and schist), granitic gneiss, and Sandia granite are well exposed in the roadcuts on the drive to the picnic area. The basement rocks are highly fractured and faulted along here, because of our proximity to the Sandia fault and related faults that separate the mountains from the basin. (0.1)

4.1 Small outcrops of reddish Triassic sandstone and mudstone (and possibly Jurassic Morrison beds) are exposed in the arroyo at 9:00. These very contorted beds are probably fault-bounded wedges (horsts) caught within a fault zone. (0.4)

4.5 Sandia Peak Tramway terminal building is at 11:30. In curve ahead, roadcuts are in bouldery fan deposits of late Pleistocene age. (0.4)

4.9 Stop Sign. Turn left onto Tramway Road. Sandia Heights residential area is on left. The road is built on a thin cover of granitic alluvium with scattered granitic boulders. Granitic bedrock is exposed in the arroyos on both sides of the road. (0.8)

5.7 Stop sign at tramway toll station. Pay parking fee and continue to parking lots at tramway terminal. (If you are not planning on riding the tram today, turn around and rejoin the road log at mile 0.0.) (0.1)

5.8 STOP 1—Sandia Crest via the world's longest tramway (5.4 aerial miles).

Park your vehicle in the lot. Walk around the terminal area to observe the giant, rounded granite boulders that have mostly weathered in place. Notice the large, rectangular crystals of feldspar that dominate the rock. The dark matrix consists mainly of biotite mica, with some small grains of magnetite. Quartz is the grayish, glassy mineral among the feldspars and biotite. The coarse sandy deposit on the ground is called "grus" (German for grit or fine gravel). It is a distinctive product of in-place granular disintegration of granite.

Sandia Peak tram at sunset, approaching the Sandia Crest.

The 15-minute tramway ride affords spectacular views of the steep escarpment, the Great Unconformity, and the surrounding countryside. Tickets can be purchased at the terminal. Restaurants are located at the base and at the top of the tram. The tramway ride begins in the Sonoran life zone at about 6,500 feet and ends at the top in the Hudsonian life zone at over 10,000 feet. This is equivalent to the change in vegetation near sea level from Mexico to northern Canada. Temperatures on top are typically 10–15° F cooler than the bottom. The crest has excellent hiking trails and a ski lift that operates year round

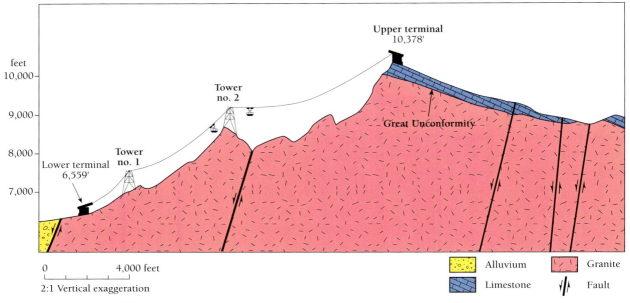

Geologic cross section along the Sandia Peak Tramway route.

(check for operating times).

Much of the west side of the Sandia Mountains was designated the Sandia Mountain Wilderness in 1978 by Congress. This designation protects the wilderness character of this rugged land. Permits are not required to visit, but the maximum group size is 25, and no motorized vehicles or bicycles are allowed. The wilderness is split into north and south halves by the tramway; a strip of non-wilderness, about 1,500 feet wide, runs beneath the cables.

THE TRAM RIDE The thrill of the comfortable ride to the top of the mountain is due largely to the sheer granite cliffs and pinnacles that rise close to the tramway, and the feeling of dangling so high above the ground. In less than 15 minutes the car rises from 6,559 feet to 10,378 feet, in a horizontal distance of 2.7 miles, traveling at a speed of 12 miles per hour. During the ascent to the first tower (2 minutes), look at the hill to the north, armored with great boulders formed by weathering of the sharp edges and corners of large jointed blocks of granite. The residual, spheroidal boulders give way upward to angular blocks and sharp angular outcrops.

As you approach the second tower, left of the car near the top of the ridge, a light-colored smooth surface marks a landslide scar where a lightning bolt knocked down some 4,000 tons of rock in 1936.

If you look out of the front or back of the car at the suspending cables, you will see orange-colored numbered markers (slat carriers) that are spaced about a minute apart. There are 11 such markers along the route. The second tower, at 8,750 feet, stands on a sharp, precipitous ridge. The cable now levels off as it swings across Baca Canyon tributaries and ascends to the crest in a single span 7,700 feet in length. Over the deepest canyon (at about marker 10), it hangs 1,000 feet above the ground.

To the southwest you can see the narrow mouth of Baca Canyon. The two large canyons south of this are Pino and Bear, site of the Elena Gallegos Picnic Area. North of the car, about midway on the span, you pass two imposing granite buttresses, one between markers 7 and 8 and the other at 9. These buttress-like spurs were formed by huge slabs of rock spalling away along great fractures (called joints). The distance from the car to the first buttress is about 300 feet; to the second, about 200 feet.

Past the second buttress you rise rapidly toward the crest, and soon the Great Unconformity comes into view: The sharp crags of the granite give way to smooth, brush-covered slopes in the layered sedimentary rocks. More than one billion years elapsed between the time the granite cooled and crystallized deep in Earth and the time it was covered by the Pennsylvanian sea.

Look closely at the Pennsylvanian sedimentary rocks ahead. Notice that they can be divided into four major cliffs (of limestone) that are separated by thin, more easily eroded layers. The eroded layers are sandstones and shales. The sandstones represent the shallowest water deposition, the shales represent the deepest water sedimentary environment, and the limestones represent the intermediate depth off-shore environment of the ancient seas. These records have been correlated with similar sea-level changes throughout North America, probably due to increases and decreases in the sizes of continental glaciers in the southern continents.

If you look out of the north side of the car between markers 10 and 11 along the base of the cliffs back toward the northwest, you may be able to make out the signs of a large fault that passes through saddles and gulches across the buttress ridges.

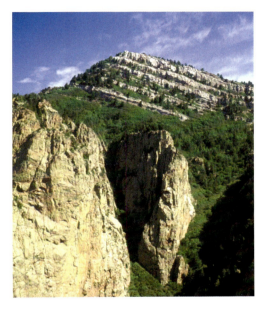

The Great Unconformity, visible to the north from the tram past the second tower.

From the top of the mountain the vista encompasses 11,000 square miles of central New Mexico. From the Summit House, which offers meals and a "top o' the mountain" view to visitors, you can walk a short distance to the upper terminal of the Sandia Peak Ski Area chairlift for a ride down the other side of the mountain. The altitude here is 300 feet lower than at Sandia Crest, which lies about a mile and a half to the north. If you have the time and endurance, the trail from here to the crest is highly recommended.

Return to the tram coach and descend to lower terminal. At the lower tram

terminal, retrace the route back to Tramway Boulevard. On the drive down, notice views of Mt. Taylor (11:00) and Cabezon Peak (1:00) to the west on a clear day. Mt. Taylor is more than 60 miles away. In the middle distance, the Albuquerque volcanoes are visible. Southward along the western edge of the basin are the rugged outlines of the Sierra Ladrones and the Magdalena Mountains.

been engineered to accommodate flash flooding that could severely damage the homes below. The floods that come racing down these channels do not consist of just water, but rather of a thick mixture of water, silt, sand, vegetation, and boulders known as a debris flow. The density of such flows is sufficient to raft large rocks and autos on top of the flows. Combined with the steep gradients of the channels,

The view to the south from the crest of the Sandias.

0.0 **Intersection of Tramway Road with Tramway Boulevard. Reset trip meter to 0.0 and turn left onto Tramway Boulevard.** For the next 8 miles, the route will cross several arroyos cut into alluvial-fan deposits. Many of these channels have

these flows are fast moving and powerful enough to destroy homes and block roads. (1.0)

1.0 Crossing Paseo del Norte Road. Note the diversion channel along the west side of

ALBUQUERQUE'S FLOODS AND DRAINS

Albuquerque is subject to sudden and severe floods, due to a combination of steep slopes, sparse vegetation, and summer monsoons, which can dump 2–3 inches of rain in a few hours. The Albuquerque Metropolitan Arroyo Flood Control Authority (AMAFCA) is the agency that attempts to reduce the personal injuries and property damage associated with floods. AMAFCA estimates that dangerous floods occur within the city about 12 times per year. Many of these floods are far removed from the rainstorm: a cloudburst in the Sandia Mountains can send a raging torrent down an arroyo or channel miles from the storm. Such flash floods can occur from June to September, but most occur in late summer when warm moist air from the Gulf of Mexico begins to ascend the slopes of the mountains. Under these conditions, towering thunderheads will develop rapidly, and 2 or more inches of rain can fall in an hour. Albuquerque also receives high-density rains over the city itself.

Flood danger is increased by urbanization of the desert landscape. For example, houses and pavement do not absorb rainfall, so the amount of runoff is radically increased, and the rainwater will quickly choke the storm drains, resulting in curb-deep runoff down city streets.

Floodwaters that move rapidly down the Sandia Mountains and its alluvial fans can pick up large concentrations of sediment, which make the waters much more destructive. These flows can damage structures that commonly line arroyos and channels. Parts of Albuquerque were developed without consideration of drainage, or before regulations were enacted. Development within arroyos, improper road grades, and undersized and poorly engineered culverts all lead to property damage during major floods.

AMAFCA, which was created by the state legislature in 1963, builds, operates, and maintains most of the dams and larger channels here on the east side of the Albuquerque Basin. The city owns most of the smaller drains and channels, many of which empty into the large AMAFCA structures. To date, AMAFCA has built 30 dams, over 50 miles of channels, and 6 miles of underground culverts, many of which we see on this trip.

the road. This channel empties into South Domingo Baca Arroyo just ahead. (0.3)

1.3 South Domingo Baca Arroyo. Note the large boulders in this arroyo and in the next several arroyos. As we proceed south on Tramway Boulevard, arroyo boulders become more abundant and larger as the road comes closer to the mountain front. (0.8)

2.1 Pino detention dam at 2:30 on Pino Arroyo. The detention dams that we will see along here are designed to capture debris flows while allowing some of the water to escape. The remaining water within the detention basin is allowed to percolate underground to help recharge the aquifer. These dams and other flood-control structures are operated by the Albuquerque Metropolitan Arroyo Flood Control Authority. The detention basins require periodic maintenance to remove sediment. (0.1)

2.2 Turn left at the entrance to Albert G. Simms City Park and Elena Gallegos Picnic Area. Created in 1984, this city park consists of 640 acres of open space on the west side of the Sandia Mountains. Facilities include picnic shelters, trails (including the Pino trailhead), and a wildlife blind situated at a pond. There is a nominal parking fee. The city sponsors nature hikes and educational programs during the summer. For more information call the Open Space Division at (505) 873-6620. Ahead, on the left, notice the large and expensive homes, some of which are built within Pino Arroyo. (1.4)

3.6 Stop at entrance booth and pay user fee, and ask for a map of the park that shows hiking trails. At the fork in the road ahead, turn right to access trailheads and main picnic area. (0.4)

4.0 Restrooms and main parking area. From here several hikes are available. It's an easy hike to the north across Pino Arroyo to visit the Kiwanis Observation Wildlife Blind at the wildlife pond. Longer

The wildlife blind, approaching the pond at Albert G. Simms City Park.

THE ELENA GALLEGOS GRANT

The Elena Gallegos Grant, the largest land grant in Albuquerque's North Valley, originally consisted of 70,000 acres and was issued to Diego Montoya in 1694. In 1716 the grant was transferred to Elena Gallegos de Gurulé; it has since been known as the Elena Gallegos Grant. As time passed, the grant was divided into narrow rectangular strips that fronted the river, some of the strips going to heirs, and some sold to non-family members. The land east of the irrigated strips up into the mountains was a commons used by everyone on the grant for grazing and woodcutting. A few of the small communities that formed inside the grant along the Camino Real are distinct villages today, but others have been enveloped by Albuquerque. In the 1920s a parcel of 35,000 acres of the grant was bought for back taxes by Albert G. Simms. Having no heirs, Simms left the land to the Albuquerque Academy when he died. They in turn sold 7,640 acres to the City of Albuquerque, when the citizens of Albuquerque voted for a quarter cent sales tax to keep the land undeveloped. The city traded 7,000 acres to the federal government to add to the Sandia Mountain Wilderness, and used the remaining 640 acres for the Albert G. Simms City Park.

treks will take you into the canyons, up the mountain, or along the mountain front. Notice the two blue tram towers visible on the mountainside to the north.

The Pino embayment is dominated by a large alluvial fan, middle Pleistocene in age. Progressively younger remnants of terraces are inset into the drainage channels of this fan. These Quaternary deposits are very thin, and the Precambrian Sandia granite is locally visible below. This is important for understanding the aquifer here, because it means that an alluvial aquifer does not exist. The presence of features such as springs and vegetation stands suggests that ground water is probably migrating through granite bedrock along faults and fracture zones. Retrace route to Tramway Boulevard and reset trip meter to 0.0.

0.0 Intersection with Tramway Boulevard. Turn left and continue south. (0.3)

0.3 Academy Road. Tramway diversion channel on right. In areas just to the west and southwest, young (Holocene) arroyos have locally built low alluvial fans. When initially deposited, these fans are not well compacted. Adding water to these deposits at a later time causes them to consolidate. If a building or road is constructed on top of these compactible deposits and water is added (for example, by watering lawns), structures can be damaged as the ground cracks and settles. Buildings and sections of roads in Albuquerque have been severely damaged by these collapsible soils. (0.7)

1.0 Main channel of Bear Canyon Arroyo. The John B. Robert detention dam is visible a half mile down the channel. A small Anasazi site at the head of Bear Canyon Arroyo was partially excavated by Albuquerque Academy students under the supervision of Gordon Page in 1983. The U-shaped structure was built around a plaza opening to the south. Although the inhabitants seem to have packed up nearly every usable item when they moved out for the last time, some clues to their everyday lives were left behind. Two sets of holes that were probably the sockets for looms were found in the floor of one room. Part of a mountain sheep skull was found embedded in the ash of a hearth in one room. Bear Canyon Pueblo was apparently started between 1150 and 1200, lived in for a short time, and then abandoned for most of the 1200s. In the late l200s or early 1300s, people returned, added new rooms, and stayed until about 1425 before they left the site forever. Other sites in the Southwest show this same pattern of abandonment and reoccupation—perhaps a response to changes in rainfall, always the most critical environmental factor for farming people. (1.8)

2.8 Candelaria Road. Begin ascending north slope of Embudo Canyon alluvial fan. (0.5)

3.3 Menaul Boulevard. Berm along west side of road is designed to provide some flood protection to homes. (0.6)

3.9 Crossing over Embudo Arroyo. Mouth of Embudo Canyon in Sandia granite at 10:00. On the afternoon of July 9, 1988, 7 inches of rainfall produced flash floods here and in Tijeras Canyon. Flood waters were up to 7 feet deep, depositing large boulders and several feet of sediment on city streets and 200 homes. The storm resulted in one death and $3,000,000 in damage. (0.4)

4.3 Indian School Road. Route crosses high axial part of Embudo Canyon fan. Four Hills are visible in the near distance straight ahead, and the Manzano Mountains in the far distance at 11:30. The Four Hills are part of Manzano Base of the Sandia Military Reservation. (0.5)

4.8 Lomas Boulevard. Route descends south slope of Embudo Canyon fan. Tramway diversion channel on left carries water southward to empty into west-flowing channel along I–40 and then into the Rio Grande. (0.9)

5.7 Cloudview/Encantado Intersection. Right lane to enter I–40. As we drive west on I–40, Kirtland Air force Base and Albuquerque International Sunport are visible on the mesa to the southwest.

THE NORTHEAST HEIGHTS

We are in the far corner of what is known as the Northeast Heights. As Albuquerque grew during the last 50 years, it was presented with several barriers: Kirtland Air Force Base and Isleta Pueblo to the south, Sandia Pueblo to the north, and the Rio Grande to the west. Initially the city pushed eastward toward the Sandia Mountains, causing rapid development of the Northeast Heights. The Heights offered rapid road access to the major job centers at Kirtland, Sandia National Laboratories, government offices, and the University of New Mexico, and the city could easily extend its utility and road infrastructure. But the main draw was the opportunity to live, at moderate cost, near the base of the Sandia Mountains in a modern home with a fabulous view westward.

Growth was slow in the 1920s because the city's water system was privately owned, and the owners were unwilling to expand eastward. Once the city purchased the water system, expansion to the northeast occurred rapidly. By the 1940s developers and merchants began to move toward the mountains, where land was cheap and abundant. Between 1946 and 1950 Albuquerque's land area tripled, mostly in the northeast. As the city's economy shifted to the military-science industry with the development of Kirtland Field, the demand for cheap housing expanded. Although much of present Albuquerque had been annexed by 1959, the area east of Wyoming Boulevard lay vacant until the explosive growth of the last 30 to 40 years.

Development posed new problems: The mountain foothill canyons and arroyos are an important component of the natural plumbing system that joins the mountain watersheds with the basin, yet much of the development focused on the traditional rectangular grid pattern of streets and homes. This resulted in the threats of destructive flood damage and contamination of surface and ground water. Had road grids followed the natural contours, erosion would have been minimized, and the residents would have had more scenic views along their streets.

SCENIC TRIP TWO - (62.8 MILES)
To the Crest of the Sandias & Beyond

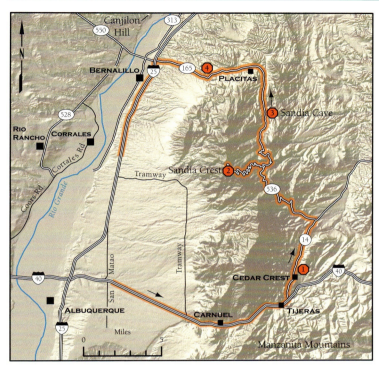

This trip focuses on the geologic and cultural histories of the Sandia Mountains, Albuquerque's eastern rampart. Highlights of the trip include Tijeras Canyon, Sandia Crest with its awesome 360° panoramic view of New Mexico, and Sandia Cave, one of the most significant Paleo-Indian sites in North America. There are many picnic areas and hiking trails along the route. As the trip winds up the east slope of the Sandia Mountains, we visit several life zones, from semiarid savannah to thick forests of spruce, fir, and aspen.

The tour begins at the interchange of San Mateo Boulevard on I–40 and ascends eastward into Tijeras Canyon. At the village of Tijeras, we turn north onto NM–14 and then west on NM–536, the Sandia Crest National Scenic Byway. From the crest, the tour turns north onto NM–165, which passes Sandia Cave en route to Placitas. The first 7.4 miles of NM–165 is a graded gravel road that is easily accessible by car, but the road can be treacherous when wet. The return route to Albuquerque is via I–25.

The tour can be driven in 3–4 hours, but we recommend packing a picnic lunch and taking a full day. In winter NM–165 may be closed due to snow, although the crest road typically remains open.

Begin the trip log at the San Mateo Boulevard I–40 overpass, headed east on I–40.

0.0 Passing under San Mateo Boulevard. The concrete-lined channel between the east- and west-bound lanes is part of an extensive flood-control system designed to prevent flooding in the city during severe thunderstorms. One unfortunate consequence of this system is that runoff is channeled directly into the Rio Grande, rather than allowed to percolate underground to recharge the aquifer or add to the soil moisture.

I–40 ascends the western slope of the Sandia Mountains along a topographic low between the large alluvial fan of Tijeras Creek to the south and coalesced alluvial fans to the north. Urbanization has made it difficult to recognize these fans. The area from here to the mountain front is covered with boulder, alluvial, and debris-flow deposits of Holocene and late Pleistocene age (less than 130,000 years old) derived from the Sandia Mountains. These coarse sediments are less than 100 feet thick. However, the underlying late Oligocene–middle Pleistocene Santa Fe Group basin fill (30–1 million years) is at least several thousand feet thick. (2.3)

Spheroidal weathering of granite creates these boulders on the lower slopes of the Sandias.

2.3 Passing under Wyoming Boulevard. The bold Sandia Mountain escarpment is prominent from 9:00–12:00. Visible on Sandia Crest, the highest peak at 10,678 feet, are many antennae belonging to local radio and TV stations. The Manzanita Mountains are the low peaks from 1:30–2:20. Tijeras Canyon is the low region at 1:00 separating the Sandia Mountains from the Manzanita Mountains.

In this area, the interstate is built on the old Tijeras alluvial fan deposited by Tijeras Creek. From the mouth of Tijeras Canyon, the fan extends westward to the center of town, sloping to a former position of the Rio Grande. The University of New Mexico campus and surroundings are built on this deposit. Tijeras Creek today flows within a deep canyon to the south. (1.5)

3.8 The four aligned hills at 2:00 are the Four Hills, part of Manzano Base of the Sandia Military Reservation. As the primary underground storage area for the U.S. nuclear arsenal, access has been restricted for nearly 50 years. (1.8)

5.6 Passing over Tramway Boulevard. Deeply incised Tijeras Arroyo is at 3:00. Sandia granite is exposed in the roadcuts ahead and in the surrounding hills. (0.4)

6.0 The boulders on the left slopes formed in place by weathering of the edges of jointed blocks of granite—a process known as spheroidal weathering. These boulders grade into the less weathered granite in roadcuts over the next 3 miles. The Sandia granite is a large pluton that consists of coarse-grained, pinkish granite with noticeably large crystals of potassium

> ### VILLAGE OF CARNUEL
> This village was one of several established on the Cañon de Carnuel Grant, which was laid out on both sides of San Antonio Creek and Tijeras Creek in 1763 as a buffer against the Faraon Apaches, who had found Tijeras Canyon an ideal staging area for raids on Albuquerque. The first town, San Miguel de Laredo, quickly became a center for trade with the friendly Carlana Apaches, but the residents fled to Albuquerque after a raid by Gila Apaches in October 1770. When they refused to return to the town, the houses and barns were razed in May 1771. The area was not resettled until February 1819, when a new grant was issued. In spite of continued attacks, most residents returned after the raids. Several new towns were established over the years (Tijeras, Cañoncito, Ranchitos, San Antonio de Padua, and Carnuel), an irrigation system was installed (parts of which are still in use), and a church was built at San Antonio in 1830. After the mines at Golden and San Pedro opened in the mid-1800s, the villages grew, and the new villages of Primera Agua, Cedro, Ojo del Sabino, Juan Tomás, Zamora, Tecolote (Gutierrez), and Sedillo were settled.

feldspar. The pluton extends along the western face of the Sandia Mountains from here northward to near Placitas, a distance of about 15 miles. (0.4)

6.4 Leaving Albuquerque city limits and entering Tijeras Canyon. Although the course of Tijeras Creek has shifted through time in the Albuquerque Basin, the course through the bedrock canyon here in the mountains has probably remained unchanged for over a million years. This canyon between the Sandia and Manzanita/Manzano Mountains was used for centuries by Indians, Spanish explorers, traders, trappers, and the pioneers heading west. The highest peaks in both ranges are over 10,000 feet, whereas this pass tops out at about 7,000 feet. (0.2)

6.6 Roadcut on left, draped with wire mesh, in Sandia granite is cut by several high-angle faults. Note the fresh granite on the left that was blasted open during road work. The transition upward from fresh, unweathered, gray granite to darker, more weathered boulders at the top of the cut is visible here. (0.4)

7.0 Passing Exit 170. Continue east on I–40. Ridge on the skyline from 12:00–2:00 is composed of Precambrian rocks capped by layered Pennsylvanian sedimentary rocks (limestone, shale, and sandstone). These strata dip east, so as we climb eastward up the canyon, we will see them ahead at road level. (0.6)

7.6 Village of Carnuel on right. Trees here are mostly juniper (Upper Sonoran life zone). Native mountain cottonwoods grow along Tijeras Creek. To the north on the high mountain crest, layered sedimentary beds (limestone, shale, and sandstone) of Pennsylvanian age (300 million years old) cap the Sandia granite (1.4 billion years old). Notice that the round weathered granite boulders give way at higher altitude to bold, angular, craggy, and pinnacled outcrops. Ahead, the gray Sandia granite grades into the pinkish Cibola granite (also known as Cibola gneiss). The

Cibola has been highly deformed, or sheared, under very high temperatures and pressures. The shearing may have occurred when the granite melt was cooling and solidifying deep underground. One recent theory states that this zone of shearing represents the bottom of the ancient Sandia granite magma chamber. (1.6)

9.2 The rocks in the roadcuts are extensively fractured Cibola granite. The fracturing is related to our proximity to the Tijeras fault zone, a major crustal zone of deformation just ahead. (0.4)

9.6 The ridges on the left and right are held up by a backbone of light-gray Precambrian quartzite. Quartzite is a metamorphic rock that originated as a quartz sandstone. During metamorphism the quartz grains recrystallize. The quartzite ridge stands out because it is more resistant to erosion than the surrounding rock, a process known as differential erosion. Along the slopes, the quartzite is mostly buried by a mixture of weathered bedrock and soil. (0.3)

9.9 Crossing over old Route 66. Across the old highway on the right side of the road are gravel terraces on both sides of Tijeras Creek. These terraces are former canyon bottoms created as Tijeras Creek cut and filled its channel. Each terrace level represents a stable period during the life of the creek. Piñon trees intermix with junipers.

Notice how straight the canyon and road are for the mile ahead. We are crossing into (and running parallel to) the northeast-trending Tijeras fault zone, buried under the creek alluvium on the right. The fractured rocks of the fault zone erode more easily than unfractured rocks, so Tijeras Creek has followed the fault along this part of Tijeras Canyon. Across canyon on the right, note the small alluvial fans formed from debris that has eroded from the hillslopes. (0.6)

10.5 A greenish, 4–5-foot-thick lampro-

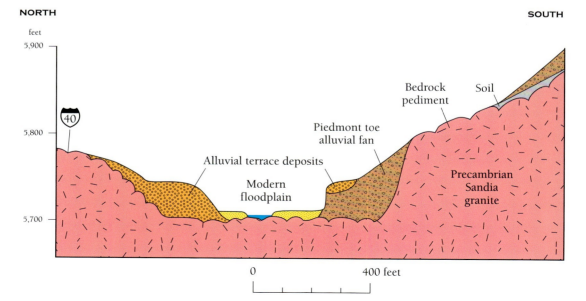

Cross section along Tijeras Arroyo.

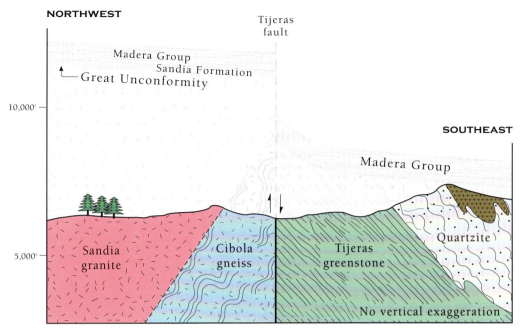

Cross section along the Tijeras fault.

phyre dike is visible in the roadcut on the left. Lamprophyre is a general term for a group of dark-colored, shallowly intruded igneous rocks that contain abundant mafic minerals, such asbiotite, hornblende, and pyroxene. These dikes may have been intruded along fractures or faults, and are probably Tertiary in age.

The dark-green and brownish rocks on the canyon slopes to the right are part of the Tijeras greenstone, a group of volcanic rocks that erupted onto a Precambrian landscape nearly 2 billion years ago. After eruption, they were deeply buried and metamorphosed. The greenstone lies under Pennsylvanian sedimentary rocks of the Madera Group. High to the north of the road, a thin cap of the same formation lies on top of Sandia granite. The greenstone and sedimentary beds south of Tijeras Canyon have dropped several hundred feet compared to the granites and sedimentary rocks north of the canyon. This displacement occurred on the Tijeras fault zone. The fault roughly follows the canyon bottom from the saddle behind us to just left of the green and white houses on the hillside in line with the road ahead. (0.3)

10.8 The Tijeras fault is exposed in the roadcut to the left where unconsolidated gravels overlie greenstone. The highway and canyon diverge from the fault here, but we will recross the fault zone ahead on NM–14 north of Tijeras. (0.2)

A greenish lamprophyre dike visible at 10.5.

11.0 Passing the site of the old Seven Springs Cafe on right. The cafe was on Route 66. Greenstone is exposed in roadcuts on the left. Limestone beds of the Madera Group crop out at the bottom of the canyon on your right and again in the roadcuts ahead on your left. As you pass, notice the gray and red shale layers interbedded with the gray limestone. These beds were warped by movement along the Tijeras fault, which was probably active in both the Laramide orogeny (about 65 million years ago) and more recently during formation of the Rio Grande rift. Prepare to exit right onto NM–14/337. (0.5)

11.5 Roadcuts ahead are coated with gunite, a cement-like material used to slow erosion. Roadcuts are commonly the best exposures that geologists can find.

Unfortunately, the application of gunite makes it difficult or impossible to study these exposures.

In 1954 University of New Mexico geology students discovered a splendid assemblage of Pennsylvanian fossil plants in these Madera Group roadcuts. The fossils were preserved as brown-stained impressions in a green mudstone. The specimens, including seed ferns and lycopods, are now at the geology museum at the University of New Mexico. Carlito Spring is located up a canyon to the left. Notice the brick-red ground straight ahead past the highway. We are about to cross from the Pennsylvanian marine limestones into Permian terrestrial sedimentary rocks that have this characteristic brick-red color. (0.6)

12.1 Exit I–40 at Exit 175 onto NM–14/337. Stay left on NM–14 to Cedar Crest. (0.8)

12.9 Road forks ahead. Keep left on NM–14 and continue through underpass.

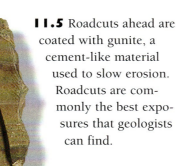

KINNEY BRICK QUARRY

The Kinney Brick Company Quarry is located 8 miles to the south in the Manzanita Mountains. This fossil locality, discovered in 1961, contains a remarkably large and diverse assemblage of fossils in the Madera Group, preserved in a 90-foot-thick section of limestone, shale, and sandstone. Few Paleozoic fossil localities in the world contain such an extraordinary variety of plants and nonmarine animals. The most remarkable plant fossil is a large leaf (nearly 3 feet long) of a ginkgophyte, a subclass of gymnosperms, a group that includes modern pine, fir, spruce, ginkgo, and cycad. The animals represented in the quarry include insects, millipedes, fishes (above), amphibians, and eurypterids (extinct arthropods, closely related to scorpions, that swam and hunted in these ancient lagoons).

The depositional environment here is a river delta that advanced into a shallow sea. The remarkable preservation of the delicate plants and animals occurred in sheltered environments that periodically became oxygen-depleted, thus retarding the decomposition of organic matter. The quarry is not open to the public.

NM–14 is the Turquoise Trail, a National Scenic Byway. The road right leads to the village of Tijeras, established in 1856. In 1973 Tijeras became the first east-mountain community to incorporate, in order to plan its own future. The limestone and shale in the ridges to the south contain assorted fossils.

The reddish-brown beds in the roadcuts are mudstone and sandstone of the Abo Formation of Permian age (280 million years). They lie on top of the Madera Group and are thought to have been deposited by rivers on a vast floodplain, perhaps under tropical conditions. The Abo Formation contains no marine fossil shells; however, in places plant remains and footprints of reptiles have been found. In contrast, the Madera Group contains abundant fossils of such marine creatures as corals, crinoids, brachiopods, and snails. (0.1)

13.0 Crossing under Interstate 40. Red siltstone and sandstone of Permian age Abo Formation is exposed along the road. The spectacular red pigment of the Abo Formation is due in part to the presence of red-to-orange feldspar grains, but most of the effect is due to brilliantly colored hematite (iron oxide) that has coated the sand grains and stained the clays. (0.2)

13.2 Crossing the Gutierrez fault, a southern branch of the Tijeras fault system. Along here we travel across red Permian rocks onto steeply dipping Cretaceous rocks of the Mesaverde Group. (0.5)

13.8 The easily eroded Mancos Shale underlies the road and forms this valley. Rocks of the Mesaverde Group crop out to the right across the stream valley. The Mancos consists of dark shale deposited as deep ocean muds during the last great transgression of the sea in Cretaceous time. The overlying Mesaverde Group sandstones and siltstones represent river deposits that flowed into the retreating ocean basins about 80 million years ago. (0.2)

14.0 MP 1. We are now 6,600 feet above sea level. The vegetation at this altitude consists mostly of piñon and juniper interspersed with some cholla cactus, ground spruce, and scrub oak. (0.4)

14.4 The road crosses over a buried fault. We know this because the roadcuts ahead are grayish sandstone and shale of the Mesaverde Group. This large-displacement fault has dropped these Cretaceous beds about 2,000 feet against the Permian red beds to the south. For several miles the road crosses Cretaceous rocks until it recrosses the Tijeras fault, which brings up the older rocks again. The downdropped wedge is informally called the Tijeras graben. The geology along and within the graben is extremely complex due to the large number of faults, folds, and stratigraphic units. (0.4)

14.8 The village of San Antonio, an early trading point along an oxcart route from Albuquerque to Santa Fe. The village was established after the second Cañon de Carnuel Grant was issued in 1819. This village is one of many small communities scattered throughout the Sandia and Manzano Mountains that are occupied by descendants of the original Spanish settlers. (0.2)

15.0 MP 2. To the right across the stream valley rocks of the Mesaverde Group are bent into a syncline (folded to form a valley) that is probably associated with the Tijeras fault. Several such folds are found in this area. (0.2)

Rocks of Mesozoic age are exposed in the outcrop on the far side of the parking lot at Stop 1. The light-colored strata on the bottom are Todilto Formation, most of the rocks above are Summerville Formation.

15.2 Entering Cedar Crest. A man named Carl Webb was one of the founders of Cedar Crest. In 1922, 23-year-old Webb left Albuquerque to live here in the mountains. He built some rental cabins, opened the Webb Trading Post (later to become the Cedar Crest Trading Post), and operated the Cedar Crest Post Office. In 1955 he sold his properties and moved away. Although he returned to Albuquerque in the 1970s, he died in Pennsylvania in 1992 at the age of 93. (0.3)

15.5 The reddish gravels in the roadcut on the left represent Quaternary alluvial-fan deposits that were shed from weathered Permian and Triassic rocks in the Sandia Mountains to the west. The canyon on the right, which the road now follows, is Arroyo San Antonio.

Just ahead, the buried Tijeras fault crosses the road diagonally from the left and continues up the valley on the right side of the road. Here the right, or east, side of the fault dropped about 1,500 feet and forms part of the Tijeras Basin, or graben. Ponderosa pines begin to appear prominently here on the hills and along the road. (0.7)

16.2 The roadcut on the right is a great exposure of fractured and faulted Jurassic Morrison Formation, a unit famed for its dinosaur fossils. Although this exposure lacks dinosaurs, it does contain numerous fault slickensides (polished and striated surface produced by movement along a fault), probably related to the Tijeras fault, which trends up Arroyo San Antonio just to our right. (0.1)

16.3 Cedar Crest Post Office on right. (0.2)

16.5 STOP 1. Pull into the parking lot of the Mountain Christian Church on the right. Road left leads to the village of Cañoncito and to Cole Springs Picnic Area (3 miles).

The small quarry in the hill exposed in the parking lot once produced gypsum for the Ideal Cement Company. This deposit proved too difficult and too small to mine, and the company now gets its supply from a larger deposit at the White Mesa mine near San Ysidro. The quarry also exposes some of the Mesozoic section. The white-mottled gypsum unit, the Todilto Formation, is visible at road level.

The finely layered red, yellow, and gray rocks above the gypsum are the Summerville Formation, with sandstones of the Morrison Formation on the very top.

The valley is eroded in red-brown Triassic shale of the Chinle Group. The ridge behind the church is Triassic shale of the Santa Rosa Formation, which lies beneath the Chinle Group and extends at depth beneath the road. All of the formations on this side of the Sandia Mountains have been tilted to the east. The Santa Rosa Formation here was mined during the 1930s for flagstone used in walks and for floors in several Albuquerque buildings.

Note the high crest of the Sandias to the west and the tree-covered slope, formed by erosional stripping parallel to the beds of limestone of the Madera Group.

Along the road ahead is the light-colored sandstone of the Entrada Sandstone, deposited as sand dunes in a Jurassic desert. Also exposed in the roadcut ahead is the sharp, unconformable contact between the Entrada and underlying dark-maroon, Chinle Group. The Triassic and Jurassic unconformity here represents about 30 million years of missing time. **Continue north on NM–14. (0.2)**

16.7 Reddish-brown Chinle Group shale is exposed on both sides of the road. The Chinle is well known to paleontologists as the source of New Mexico's state fossil, *Coelophysis*. (0.9)

17.6 Summit of hill. This is the major drainage divide on the east side of the mountains. San Pedro Creek, ahead, drains north from here around the mountains and empties into the Rio Grande near San Felipe Pueblo. South of this divide, Tijeras Creek and its tributaries drain the mountains to the Rio Grande south of Albuquerque. (0.8)

18.4 Road bends left. View of Ortiz Mountains straight ahead and San Pedro Mountains at 1:00–2:00. (0.5)

18.9 **Turn left onto NM–536, the road to the Sandia Crest.** NM–14 continues north along the Turquoise Trail to Golden, Madrid, Cerrillos, and Santa Fe. Trip 3 starts here. The wide valley here still lies in Triassic Chinle Group shale, although far to the southeast, the curving valley rimmed by a tree-covered ridge is Cretaceous shale exposed east of the Tijeras fault. (0.9)

19.8 Roadcuts on the right are in reddish sandstone and shale of the Abo Formation, the same Permian red beds as those around the village of Tijeras. Notice the small faults that offsets bedding ledges in the red beds. Ponderosa pines are increasingly abundant. (0.2)

20.0 MP 1. The Tinkertown Museum, just ahead on left, contains an animated, miniature, wood-carved western town and circus built and operated by Ross and Carla Ward. Their motto is "We did all this while you were watching TV." Ross Ward began carving the first figures for the General Store building in 1962. Thirty years later, he completed the entire miniature western town and circus with over 20,000 miniatures. The complex contains a remarkable assortment of collectibles from the old West. The museum collections are housed within 22 rooms whose walls are constructed from over 50,000 bottles of all shapes and colors. Museum is closed during winter months.

Just to the north was the site of the main Civilian Conservation Corps (CCC) camp of the 1940s. The CCC crews worked on picnic grounds, water systems, shelters, fences, trails, a ski lodge, and the Tijeras Ranger Station. (0.5)

20.5 Entering the Sandia Ranger District of the Cibola National Forest. Congress set this area aside as a national forest in 1907. No commercial logging has occurred since 1969 and no cattle grazing since 1951. This heavily timbered area is home to deer, bears, wild turkeys, mountain lions, and bighorn sheep. Elm trees grow along the creek on the left, and some large scrub oaks are visible ahead. More than 2 million people visit the Sandias annually, making it the most visited mountain range in the state. In this district 37,232 acres of the 100,555 acres are protected as the Sandia Mountain Wilderness. All motorized and mechanized equipment, including bicycles, are prohibited on the 117 miles of maintained hiking trails in the Wilderness. No overnight camping is allowed. (0.1)

20.6 Crossing the contact between Permian Abo Formation and underlying Pennsylvanian Madera Group, which is exposed as limestone, black shale, and sandstone beds in the roadcuts ahead on right. (0.1)

20.7 Sulphur Springs Canyon Picnic Ground, elevation 6,800 feet, contains restrooms, picnic facilities, and a trailhead. Road now passes into the Madera Group; fossils are visible in the roadcuts ahead. (0.2)

20.9 MP 2. We have just recrossed the Great Unconformity between Pennsylvanian strata and the Precambrian Sandia granite, exposed in roadcuts just ahead. This is the same geologic contact seen on the western face of the Sandia Mountains, where stratified sedimentary rim rocks rest on the massive granite cliffs. This unconformity represents a gap in the record of geologic time of more than one billion years between the 1.45-billion-year-old granite and the 320-million-year-old Sandia Formation. Fortunately for the geologist, much of the missing record is preserved in surrounding regions of New Mexico, Colorado, and in the walls of the Grand Canyon in Arizona.

The discoloration and "rotten" look of the granite beneath this unconformity results partly from recent weathering and partly from weathering on the ancient erosion surface before burial by the sedi-

The Great Unconformity, exposed at 20.9, represents over a billion years of missing geologic record.

mentary beds. A thin gravel bed overlies the granite, and marine fossils can be found 5–6 feet above the base and in scattered beds upward in the sequence. The sandstone and shale beds are typically mica rich, suggesting a mica-rich rock (such as granite) as an erosional source. Hundreds of feet of Pennsylvanian sedimentary strata overlie the Sandia granite, yet the granite is exposed here. This is because a large north-trending fault (the Barro Canyon fault) lies just a few hundred feet west of the road. This fault "broke the back" of the Sandias and raised a small piece of the Sandia Mountains block, exposing the granite.

Box elder, scrub and Gambel oak, cottonwood, some juniper, and pine trees line the canyon bottom to your left. (0.1)

21.0 Side road to the Doc Long Picnic Area, a favorite recreational and culinary spot of Albuquerqueans. Facilities include restrooms, tables, and a trailhead. CCC crews built this picnic ground in 1935. This picnic area is named for Dr. William H. Long who worked here from 1910 to the mid-1930s as a forest pathologist. He was one of the three in the country studying tree diseases. Abert's squirrels (*Sciurus aberti*) live in these ponderosa pine forests. Their ancestors were transplanted from south-central New Mexico in the 1940s. (0.4)

21.4 Hairpin turn. Many dikes crosscut the Sandia granite here. These dikes formed when fine-grained granite was injected as molten rock into older rocks, commonly along fractures. Ponderosa pines and scrub oaks dominate the flora. (0.1)

21.5 The Barro Canyon fault is visible in the roadcut on the right. Rocks west of the fault are the uppermost beds of the Madera Group, which in this area may be 1,200 feet thick. As the road turns the corner ahead and goes up Tejano (Texan) Canyon, it cuts stratigraphically down through eastward-dipping beds; in the next 0.8 mile, you descend stratigraphically through almost the entire 1,200 feet of the formation. Many fossil-bearing beds are exposed in roadcuts along this stretch. (0.3)

An 18-foot-wide dike of igneous rock cuts limestone beds in the roadcut at mile 22.2.

21.8 Excellent roadcut exposure of east-dipping limestone and shale. North slope across the canyon to the left has thick stands of white fir, whereas the sunny south slopes above the roadcuts have mostly piñon with some juniper. Ahead up the canyon along road switchback slopes, an old burn area contains only oak and firs and pines. (0.2)

22.0 MP 3. Notice that the limestone beds have shallower dips here, because we have just crossed another fault. (0.2)

22.2 A dike of igneous rock, 18 feet wide, cuts limestone beds in the roadcut on the right. Note the contact metamorphism along the contact with the country rock. The dike must occupy a fault, because the beds on opposite sides of the dike do not match. Dikes with this texture and mineralogy are called lamprophyres; they contain lots of mafic minerals (biotite, hornblende, pyroxene) in a finer-grained groundmass. (0.2)

22.4 Hairpin curve. Contrast the vegetation east and west. To the east are scrub oak, yuccas, and sparse juniper caused by overexposure on a west-facing slope. To the west, one sees mostly white fir and pine of the Canadian life zone. (0.6)

23.0 As you round the curve for a view across Tejano Canyon, notice the prominent ledges of the Madera Group. These are the same ones visible from Albuquerque on the high west face of the mountains. Uplift of another fault block and erosion bring them to view here. (0.4)

23.4 Again, the Great Unconformity contact is exposed on the left, inclined northward up the roadcut. If you examine the contact carefully, you may see several rounded, residual boulders similar to those in Tijeras Canyon. The sediments covered these weathered boulders on an ancient erosional surface. The exposures here, and even more so in the old roadcuts 100 feet above, attest vividly to the fact that intense weathering occurred here 300 million years ago in Early Pennsylvanian time. These same sediments overlie the blocky granite outcrops across Tejano Canyon. This slope provides a good example of the generally poor exposures from which geologists must trace contacts and faults (no wonder geologists love roadcuts!). (0.5)

23.9 MP 5. Limestone ledges on the right dip west here instead of east because of having been dragged along a fault. The elevation here is 8,000 feet, and we are now in the mixed conifer vegetation zone. (0.6)

24.5 Tree Spring Trail on left joins the Crest Trail about a mile south of the tram

SANDIA CREST NATIONAL SCENIC BYWAY

CCC camp in Sandia Park, 1933-40.

This 13-mile road gains 3,828 feet in elevation, an average of nearly 300 feet per mile. The temperature can easily change by 20° F from here to the crest. The road began as a series of unofficial wood cutting trails and logging skid trails. In the 1930s the trails were linked by a road from Balsam Glade to the crest, with assistance from the Civilian Conservation Corps. Between 1974 and 1982 the highway department widened and reconstructed the road from Sandia Ski Area to the crest, eliminating many dangerous curves. The road is now paved but can be slippery when wet or icy. Recreational vehicles and trailers should not ascend past the Doc Long Picnic Area.

station. The spring has been captured to provide a reliable water source for wildlife. Look for black bear, mule deer, red squirrels, Steller's jays, and other creatures around here. We are now in the Canadian zone; the Douglas-fir, white fir, and quaking aspen here at 8,480 feet can be found at sea level in Canada. (0.4)

24.9 MP 6. Just ahead on the right, Dry Camp Picnic Area contains restrooms and picnic tables. The purplish-brown rocks on the left are the highest beds of the Madera Group. Just east of here is a long-abandoned mine adit (tunnel) on Tecolote Peak. (0.9)

25.8 Sandia Peak Winter Sports Area. The ski area is a year-round recreation attraction under a special-use permit with the forest service. Skiing started here in 1938 after a CCC crew cleared the ski slopes and built a day lodge with the timber. The present lodge was built in 1983. The Sandia Peak Ski Area offers a variety of beginners to expert downhill runs, cross-country skiing, snowboarding, snowshoeing, and innertubing. The ski season has ranged from a short season of five days in 1966–67 (before snowmaking technology), to 148 days in 1972–73. During the summer, hikers can take the chair lift to the peak. The lift unloads near the restaurant where the tramway car docks. The chair lift is 7,000 feet long and takes about 15 minutes to reach the summit. The lower part of this ski run, especially around the lodge, rests on an old landslide. (0.2)

26.0 Just ahead is a view right across La Madera Canyon and into the upper part of San Pedro Valley. Two mountains are visible across the valley to the right of the winding course of NM–14. The lower one, Monte Largo, consists of gneiss and quartzite; the higher, three-peaked one with the horizontal tree line around it, is South Mountain, an igneous intrusive mass (laccolith). San Pedro Mountain is to the left. (0.4)

26.4 Junction with NM–165. Keep left on NM–536 to the crest. Reset mileage to **0.0** Balsam Glade Picnic Area has restrooms. We are now at 8,650 feet, driving along a zone of north-trending faults.

0.0 The road is on a gentle slope that almost parallels the tilt of the sedimentary strata, although slightly less steeply, so that the beds at the beginning of the road here are perhaps 100–200 feet higher stratigraphically than those at the end of the road and on the crest. Mostly white fir here, some pine and scrub oak for the next half mile. (0.5)

0.5 MP 8. Just ahead, Capulin Spring Picnic Area, 8,800 feet, contains restrooms and picnic tables. The Capulin Snow Play Area contains specially designed inner tube runs, snow permitting. There are 13 major switchbacks between here and the crest. (0.9)

1.4 Nine Mile Picnic Area, 9,200 feet, contains restrooms and tables. Notice the abundant aspen, both old and young. The wild raspberries here entice humans and animals. Keep watch for Clark's nutcrackers, Steller's jays, flickers, chickadees, nuthatches, and Audubon's warblers. (0.5)

Crest Trail.

1.9 Thinly bedded Madera Group is exposed on the right. Trees here are mostly white fir, but several large Douglas-firs and pines grow on both sides of the road. (2.8)

4.7 Ellis Trailhead Scenic Byway on left. The service road here, which leads to Summit House and the restaurant, is not open to the public. We are now well into the Hudsonian zone (spruce–fir belt), and only the hardiest trees can survive this wet, windy, cold climate. If your timing is right, you may be treated to blooms of fairy slippers and coral root orchids. (0.7)

5.4 Hairpin curve left. The powerlines here were specially designed for birds of prey (hawks and falcons) that use the mountain range as a migrational flyway. Notice the perches on top of the poles. Also notice the gnarled and bent aspen, caused by deep snow packs. Top of Madera limestone exposed along road. (0.5)

5.9 STOP 2 - Sandia Crest, 10,678 feet above sea level. Park and walk to the overlook (a day-use fee was imposed in 1997). On December 29, 1958, a record 30 inches of snow fell here on the crest. From this vantage point, on a clear day, you can easily see 100 miles in any direction. On exceptionally clear days you can see the peak of Sierra Blanca, 135 miles south, near Ruidoso. The total area of view encompasses about 15,000 square miles.

The Sangre de Cristo Mountains to the northeast represent the southern end of the Rocky Mountains. Santa Fe is nestled along the west flank of the mountains. The Jemez Mountains to the northwest are remnants of a giant collapsed volcano known as the Valles caldera. Mount Taylor to the west is the largest preserved volcano in New Mexico. The Manzano Mountains to the south form the faulted eastern edge of the Rio Grande rift. The western edge of the rift is 30 miles to the west.

We are standing on Madera Group rocks at an elevation of over 10,000 feet, yet in the Albuquerque Basin under the city, the same rocks are present at an elevation of 15,000 feet below sea level. That is an offset of nearly 5 miles! Most of the faults responsible for this enormous crustal rupture lie buried under alluvium along the base of the Sandia Mountains.

The view to the south from Sandia Crest.

Several hiking opportunities are available at the crest.

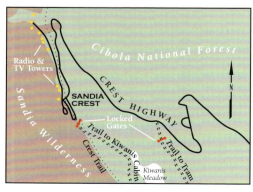

The slope of Pennsylvanian rock below us consists of four major cliffs of limestone separated by thin sandstones and shales. The sandstones represent the shallowest water deposition, the shales represent the deepest water sedimentary environment, and the limestones represent the intermediate depth off-shore environment of the ancient seas. Each of these major limestone cliffs can in turn be divided into four subunits that have been correlated with similar cycles throughout North America, due to global sea-level changes that were likely caused by the growth and retreat of continental glaciers in the southern continents. Geologic systems are complexly interconnected, and changes in one area of the globe can directly influence geologic processes in another.

Note the Sandia Crest Electronic Site to the north, a veritable forest of TV and radio towers. What began in 1945 with a state police transmitter, has grown into a major communications center for the southwest U.S. The Sandia House store here sells memorabilia and sustenance for the traveler. The La Luz Trail, a popular hiking trail that links the city with the crest, begins just below here. The Crest Trail (wheelchair accessible for the first 400 feet) begins just to the south. Return the 5.9 miles to NM–165/536 intersection. Watch downhill speed; many sharp turns.

0.0 Junction of NM–165 and NM–536. TURN LEFT into Balsam Glade Picnic Area, and bear left onto narrow dirt road (NM–165) for final leg of tour loop to Albuquerque via Placitas. Set trip meter to zero at end of pavement. The road is unpaved for the first 7.4 miles, rough in places but well graded. Passenger cars should have little problem on this road, but RVs and car trailers are discouraged. This road is closed from November 1 to May 1 because of snow. (0.2)

0.2 Hairpin turn right into Capulin Canyon. (0.6)

0.8 We have just crossed a fault that brings Precambrian rocks up to road level. Precambrian rocks exposed here and in the roadcut ahead along the hairpin curve are part of a fault-bounded block. Here the Precambrian rocks are interlayered quartzites (probably sedimentary rocks to begin with) intruded by foliated Sandia granite. Note that the bedding layers in the quartzite are vertical. (0.2)

1.0 Hairpin curve. Over the next couple of miles, as we wind down the canyon, we will cross from Precambrian to Pennsylvanian rocks several times. (0.5)

1.5 Sandstone in the roadcut on left is part of the Sandia Formation that underlies the Madera Group and covers the Great Unconformity. MP 15 is just ahead. View ahead of San Pedro Valley and the Hagan Basin. The Ortiz Mountains are in the middle distance; Santa Fe and the Sangre de Cristo Range are on the northern horizon. The ledge-capped mountain a short distance northwest is Palomas Peak, composed of Madera Group strata. (0.3)

1.8 Sharp turn left. Ahead, Sandia Formation sandstone is on inside curve; this sandstone rests on Sandia granite just

Sandia Cave (top) is in a cliff of Madera Group limestone. Sandia points (below) have been reported from several other New Mexico localities, as well.

ahead. Notice how crumbly the granite has become from weathering in the humid environment of the forest. Many more such roadcuts are ahead. Just ahead we'll pass through an area burned by a small forest fire. (0.6)

2.4 LS Ranch Road. The prominent canyon on the right is Las Huertas Canyon. Palomas Peak is to the right, with great limestone ledges of the Madera Group. The same ledges across Las Huertas Canyon are lower than the equivalent beds high on Palomas Peak. This is explained by another large, north–south fault that follows Las Huertas Canyon and has lifted the east side by nearly 1,000 feet, relative to the west. The shoulder below the big ledges on the profile of Palomas Peak marks the position of the Great Unconformity. (0.1)

2.5 Entering a private inholding on Cibola National Forest. Please stay on road. (1.0)

3.5 MP 13. The Las Huertas Picnic Ground, 7,600 feet, has a stream, restrooms, and water. It is a pleasant place for a picnic, hike, or rest, if it's not too crowded. More decomposed granite is exposed in roadcuts here. (0.6)

4.1 Granite cliffs on both sides of the road. (0.3)

4.4 The Madera Group ledges have sloped to the canyon bottom on the right and buried the granite in roadside exposures. (0.1)

4.5 We now leave Precambrian rocks behind as we continue the descent down Las Huertas Creek to Placitas. For the next several miles the bedrock consists of Pennsylvanian and Permian sedimentary rocks. (0.3)

4.8 STOP 3 – Sandia Cave. Pull into parking area on right (fee area). The cave appears as a large hole surrounded by a yellow stain within a thick layer of limestone high above the ground. The principal cave extends into the limestone hill 460 feet. A moderate but pleasant 0.5-mile trail through piñon and juniper trees leads to the cave. Exposures of Madera Group limestone are seen along the trail.

Sandia Cave is one of the best-known Paleo-Indian sites in North America. Excavations began in the 1930s and revealed several distinct levels of occupation. Below the top level, which contained Pueblo III pottery, a layer of calcium carbonate sealed the lower levels. One of these lower levels contained Folsom points, which are dated from 8,800 to 8,000 B.C. Below the Folsom level was a deposit of yellow ochre, 2–24 inches thick, which contained no artifacts.

Below the yellow ochre was a level that is now much disputed. It contained 19 specimens of a distinctive single-shouldered point (the Sandia point), apparently in association with extinct forms of horse, bison, camel, mastodon, and mammoth. Radiocarbon dates of 17,000–35,000 years ago were reported for this level. Many scholars have since raised serious questions about the true stratigraphic positions of the Sandia points with respect to the extinct animal bones, and the dates are no longer accepted by most archaeologists. Nevertheless, the Folsom level appears to be valid. The cave is one of only two Paleo-Indian sites in the Southwest that can be visited (the other is the famous Clovis site, Blackwater Draw, near Portales in eastern New Mexico). The mouth of the cave may have eroded 5–10 feet from where it was when prehistoric people occupied it, and the entrance may have been even more difficult to reach.

Continue down along NM–165. (0.5)

5.3 Crossing bridge over Las Huertas Creek. Many more bridge crossings ahead. (0.2)

5.5 Many outcrops of thin- to medium-bedded limestone and shale flank the road. The many holes, small caves, and enlarged crevices in the limestone were formed by water dissolving the calcium carbonate. (0.5)

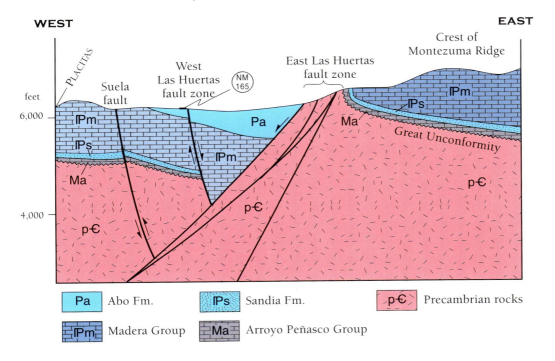

6.0 A small acequia across the stream delivers water to members of the Las Huertas Ditch Association. Archaeological investigations in 1986 found remnants of historic and prehistoric irrigation systems. One ditch apparently dates before 1680, and another may be as early as the 1400s. An extensive agricultural complex of acequias, laterals, and terraced garden plots was constructed by the settlers of San Antonio de las Huertas between 1767 and 1823, including a ditch running through the walled San José village compound. (1.3)

7.3 MP 9. Pavement begins. Montezuma Ridge is ahead and to your right. (0.5)

7.8 Reddish Abo Formation exposed on right is on the western downdropped block of the West Las Huertas fault. (0.8)

8.6 The Jemez Mountains are visible straight ahead. The valley to the right is a graben, a downdropped fault block. We are driving on the western edge of the graben, along the West Las Huertas fault, which separates Madera Group left of the road from Abo Formation right of the road. The East Las Huertas fault marks the base of Montezuma Ridge, the high ridge to the right capped by Madera Group. It separates Precambrian rocks in the lower slopes from the Abo Formation in the valley, a displacement of at least 1,000 feet. These faults are probably related to the Laramide orogeny and most likely were reactivated during development of the Rio Grande rift. (0.5)

9.1 Crossing small, north-plunging syncline in the Abo Formation. (0.4)

9.5 Crossing another major north-striking fault system (Suela fault to the south, San Francisco fault to the north) that drops Cretaceous strata to the west against Permian rocks to east. The long-abandoned coal mines of Placitas are located in the hills to the north. (0.4)

9.9 Placitas Post Office and cemetery. The flat area ahead is underlain by Quaternary gravel and sand deposits that represent former stream terraces. Several such deposits and river terraces have been identified in this region. (0.6)

10.5 Greenish beds of the Jurassic Morrison Formation in the roadcut on the left lie beneath thin, rusty, sandstone beds of the Cretaceous Dakota Sandstone. The Morrison Formation (145 million years old) is famous for its spectacular dinosaur fossils, including *Stegosaurus*, *Apatosaurus*, and *Allosaurus*, especially in Utah, Colorado, and Wyoming. It also marks the time when true birds first evolved, and flying reptiles, crocodiles, turtles, and plants were abundant. The Morrison was deposited by large, sluggish streams as alluvial-plain and river-channel sediments. The overlying Dakota Sandstone is separated from the Morrison by an unconformity that represents a gap of 35 million years. The Dakota was deposited during a major transgression that extended from the Arctic Ocean to the Gulf of Mexico. (0.2)

10.7 More Morrison Formation exposed on left. Crossing a series of buried faults. Just ahead on the right is an excellent view of the gentle eastern dip slope of the

Sandia Mountains and the prominent, cliff-forming limestone ledges that cap the range. (0.4)

11.1 Small roadcut exposure of Todilto Formation on left. Lomos Altos, the mesa to our right, consists of piedmont deposits of the Santa Fe Group. (0.1)

11.2 Crossing the Caballo fault. Unconsolidated material ahead on left is Quaternary gravel. (0.7)

11.9 Placitas Fire Station on right. The valley here overlies soft Cretaceous shales, such as the Mancos Shale, visible in some arroyos ahead. The Mancos was deposited on top of Dakota Sandstone sands as ocean-floor muds in a deeper sea and contains many marine invertebrate fossils, including the spectacular group of animals known as ammonites. (0.4)

12.3 On the skyline at 1:30 is Cabezon Peak, the eroded throat of a long-extinct volcano. The ridge on the left is mostly Morrison Formation. (0.6)

12.9 STOP 4 - Park on right at Las Placitas Historical Marker. Walk westward along the roadcut. Many geologic contacts, unconformities, and faults are beautifully exposed in this roadcut. Examine the rocks and see if you can formulate a sequence of geologic events based on cross-cutting relationships. Here is our interpretation (see photo on this page):

The easternmost exposure (closest to parking area) is the olive-green Mancos Shale. Walking westward along the roadcut, you cross into a fault-bounded slice of Morrison Formation containing multi-colored sandstone and shale. This slice (which contains smaller faults) has moved upward relative to rocks on either side. In the middle of the roadcut, you can see where the Morrison Formation is in contact with the Mancos Shale. Would you call this a fault or an erosional contact? Because of the contorted beds at the contact, we suspect that it's a fault.

Continue walking westward and cross a prominent fault that brings the Mancos Shale against much younger Santa Fe Group gravels. Examine this fault contact

Roadcut at Stop 4.

to discover rotated pebbles and fault striations. These features are characteristic of many faults. About 100 feet west, another fault cuts Santa Fe Group gravels. Overlying everything is the youngest unit in the roadcut, a Quaternary gravel deposit. Note the scoured contact at the base of the gravel deposit. These faults are part of the Ranchos fault zone, probably a Laramide-age structure reactivated during Rio Grande rifting. Careful observation coupled with some fundamental geologic principles allows us to construct a reasonable geologic history for this outcrop as follows:

1) Deposition of the Morrison Formation during Jurassic time, followed by a period of erosion or non-deposition
2) Deposition of the Mancos Shale during Late Cretaceous time
3) Deformation during the Laramide orogeny in early Tertiary time, resulting in the Morrison block being uplifted against the Mancos Shale
4) Deposition of stream gravels of the Santa Fe Group in mid-Tertiary time, with extension during Rio Grande rifting causing the Santa Fe Group to be downdropped against the Mancos Shale
5) Deposition of Quaternary gravels (probably by streams flowing westward) on top of all the other units
6) Erosion and deposition of gravels associated with modern streams

Continue driving westward. (0.4)

13.3 MP 3. As the road descends across the westward-sloping surface on Santa Fe Group sand and gravel, we cross yet another north-striking fault, the Valley View fault. In this area, the fault juxtaposes different parts of the Santa Fe Group. White Mesa at 12:30 is composed of Todilto Formation gypsum, which is being mined. Also directly ahead is the southern end of the Nacimiento Mountains. Cabezon Peak appears on the skyline to the left of the Nacimiento Mountains. At 1:30– 2:30 the southern rim of the Jemez Mountains caldera is visible. In the middle distance at 1:00–2:00 you can see the lava-capped Santa Ana Mesa across the Rio Grande. Notice how faults have offset the lava surface. (1.2)

14.5 MP 2. Descending off high alluvial surface onto younger alluvial deposits. (1.1)

15.6 Good view of the narrow green strip of bosque along the Rio Grande. Beyond the river is the higher Llano de Albuquerque surface that represents the highest level of basin filling. Just ahead,

Cabezon Peak is a volcanic neck. This vertical, pipelike intrusion represents a former volcanic vent; the cone associated with the eruption has long since eroded away. Radiometric dating indicates that this volcanic feature is 2.66 million years old.

the Albuquerque volcanoes are visible at 10:00 and Mt. Taylor at 11:30. (1.0)

16.6 I–25 interchange. CONTINUE STRAIGHT AHEAD across overpass and turn left onto the I–25 south on-ramp to Albuquerque. Reset mileage to 0.0.

0.0 On the right is Bernalillo, county seat of Sandoval County. It was originally a trading center. (0.3)

0.3 MP 242. Buff-colored Santa Fe Group sands and gravels in roadcuts ahead. The Santa Fe Group is the major basin-fill unit in the Albuquerque Basin and ranges in age from late Oligocene to middle Pleistocene. (0.6)

0.9 Water tank and gravel quarry on left. The gravel is being removed from Rio Grande terrace deposits that lie on top of the Santa Fe Group deposits. Several gravel quarries exist along the interstate.

The prominent mountain on your left, known as Rincon Ridge, forms the northern spur of the Sandias and partly encircles the Juan Tabo Recreation Area. It stands in front of the main granite escarpment of the Sandias and consists of Precambrian schist and quartzite cut by numerous igneous dikes and intruded by the Sandia granite. (1.3)

2.2 Entering the mostly undeveloped land of Sandia Pueblo. Reddish exposures that crop out in roadcuts and arroyo banks here and ahead are sand, mud, and gravel of the Santa Fe Group. Notice the Rio Grande cultivated floodplain, right, and the pinkish Santa Fe Group beds in the low slopes across the valley. (1.5)

3.7 Ladron Peak (9,716 feet) is the high point at 12:00, about 60 miles distant. The Sierra Ladrones are an uplifted block of Precambrian rock on the western side of the rift. (0.6)

4.3 MP 238. The interstate is built on terrace deposits of the Rio Grande. Note the low erosion scarp on the left paralleling the interstate. It marks the eastern edge of the inner Rio Grande valley. (2.5)

6.8 Large gravel operation on right. Spectacular assemblages of mammal fossils have been recovered from such quarries. One such discovery was of a long-extinct camel species. The find was unusual because the skeleton was complete, articulated (intact), and exceptionally well preserved. Also, the bones provided precious data on the growth of extinct camels, because the animal was only a year old when it died. The skeleton was found lying in a peculiar position, with the head and neck bent far back over the shoulders. Modern camels of the Sahara Desert that have died of starvation are usually contorted into the same position. Because the creature was found intact, it must have been buried quickly, before scavengers could rip it apart. The skeleton is now on display at the New Mexico Museum of Natural History and Science in Albuquerque. (0.5)

7.3 MP 235. Bernalillo County line.

SCENIC TRIP THREE - (58.6 MILES)
The Turquoise Trail

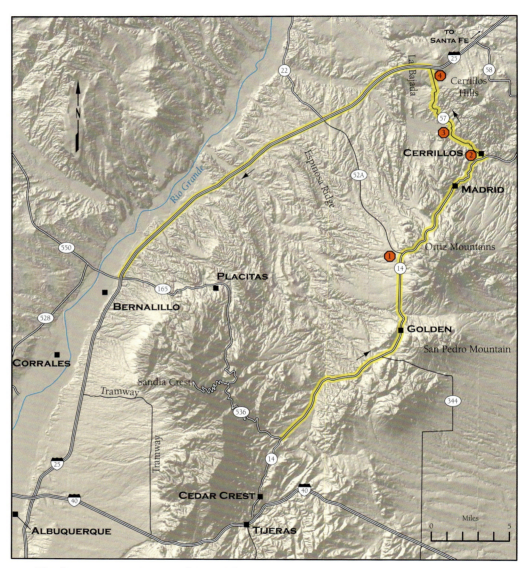

The former mining towns of San Pedro, Madrid, Golden, and Cerrillos offer the traveler a peek into New Mexico's past. Spanish miners in the Golden/San Pedro area may have undertaken the first lode mining in the western U.S. Pottery dating to the middle to late 1600s has been found in association with old smelter ruins in the

New Placers mining district. Gold was the impetus for 19th century mining at San Pedro and Golden, but the veins soon pinched out, and water sufficient for sluicing gold from the gravels was unavailable. Fortunately for the towns, copper was discovered, and the mines operated until well into the 20th century. By 1960 more than 12 million pounds of copper had been produced in the New Placers district. Madrid flourished as a coal-mining town from about 1880 to 1954. A ghost town for the next 20 years, Madrid has since undergone a revival as a mecca for artists. Madrid was world famous for its fabulous display of Christmas lights; now sculptures abound in the hills south of Madrid.

The Cerrillos mining district, now represented only by the small town of Cerrillos, is renowned worldwide for its turquoise mines. Although an estimated $2,000,000 in turquoise was taken out by Tiffany's of New York in the 1890s, it is the evidence of prehistoric mining that makes Cerrillos unique. Worked during the A.D. 800s, 900s, and 1300–1700, Cerrillos turquoise was traded not only to Chaco Canyon and other southwestern communities, but as far away as the Aztec and Mayan empires in Central America. Shafts opened by prehistoric miners using only handheld stone hammers and mauls are visible today in the Cerrillos district.

0.0 Start of trip, at intersection of NM–14 and NM–536. This spot corresponds with mile 18.9 of Trip 2. Drive north on NM–14. At milepost 6, just ahead, we are descending into the drainage of San Pedro Creek along Quaternary alluvial fans that originate in the mountains to our right. On the left are the ruins of the ancient pueblo of Paa-ko. (1.2)

1.2 Panoramic view of several local mountains: The Ortiz Mountains are visible at 11:30, and, in the far distance, the Sangre de Cristo Mountains. The San Pedro Mountains (sometimes called the Tuerto Mountains) are straight ahead on

This bowl from Paa-ko is an example of Agua Fria Glaze-on-red, the earliest glaze ware on the Rio Grande. It was made sometime between A.D. 1325 and A.D. 1425.

PAA-KO

Excavations were first conducted at Paa-ko in 1914. The first structures, built in the 1000s or 1100s, were razed when a two-story adobe structure containing more than 160 ground-floor rooms was constructed in the late 1200s. In one kiva patches of painted walls hint at what may be the earliest kiva murals known. The pueblo was abandoned about 1425 and lay vacant until 1525, when construction of huge masonry room blocks surrounding four large plazas began. This portion of the site was inhabited when the first Spanish settlers arrived in 1598. Excavations by the University of Chicago in 2000-2001 revealed the cobble foundations of a small structure that may prove to be the lost mission church of Paa-ko, long believed to exist but never identified. Extensive evidence of a Spanish smelting operation was also found, where ore was reduced to extract copper and iron. The site is now owned by the University of New Mexico.

the skyline; scars of old mines are visible. The lower ridge from 12:00 to 1:00 is Monte Largo. At 1:00, the high peaks beyond Monte Largo are South Mountain. This area has been undergoing rapid development because of its natural beauty, moderate climate, and proximity to Albuquerque. Given the limited groundwater in this area, many scientists and residents fear overpumping and polluting of the aquifers. Piñon–juniper woodlands and grasslands predominate here. (1.9)

3.1 MP 9. At 9:00, on the skyline, the cluster of antennae mark Sandia Crest at 10,678 feet (on stormy days the crest may be shrouded in clouds). The large group of cottonwood trees and shrubs just ahead on the right in the drainage marks San Pedro Spring. About a mile to the right is the Tijeras fault zone. Between here and the fault zone is a wide area of down-warped rock strata known as the San Pedro synclinorium; consequently, the Pennsylvanian and Permian section to our left is repeated in the hills to our right on opposite limbs of the fold. At 11:30 the rounded hill on the skyline is capped by Madera Group limestones. (0.8)

3.9 Notice that San Pedro Creek on the right contains an inner active channel incised in a sandy fluvial deposit. Such relationships can be used to reconstruct the recent geomorphological history of the area—the history of down-cutting and deposition by the creek, and the possible influences of factors such as climate and tectonics on this history. (0.6)

4.5 Sandoval County line. Monte Largo on the right skyline is a Precambrian-cored, fault-bounded structural high that sits within the Tijeras fault zone. The high ridge of layered rocks at 2:30– 3:00 is the Madera Group on the northwest flank of the uplift. On the skyline straight ahead are the Cerrillos Hills. (0.4)

4.9 Crossing San Pedro Creek, which flows northward and westward from here, converging with Tonque Arroyo before emptying into the Rio Grande at San Felipe Pueblo. (1.2)

6.1 MP 12. Crest of hill.(0.5)

6.6 Abo Formation red beds in roadcut on right. (0.3)

6.9 Highway turns left. Crossing the Tijeras fault again. The fault and road are nearly coincident for the next mile. Ahead, the road bends right, whereas the fault continues straight ahead across the hill at 12:00, where it separates Abo on left from Madera on right. Gravel exposed in roadcut here is dominated by light-gray-colored Proterozoic quartzite clasts derived from outcrops at Monte Largo to the right. (0.4)

7.3 Road crosses the buried Texas–New Mexico gas pipeline. The arroyo to the left contains a rare, excellent exposure of the Tijeras fault. The rocks became exposed in 1994 after a flood scoured the bedrock clean of sand. The fault zone consists of highly fractured rocks of vastly different ages that were jumbled together during faulting. (1.2)

8.5 Santa Fe County line. (0.6)

9.1 MP 15. San Pedro Mountains ahead are part of the Ortiz porphyry belt, a line of Tertiary intrusive rocks that crops out along the Tijeras fault zone. Sedimentary rocks in the San Pedro Mountains range from Triassic Chinle Group and Santa Rosa Formation on the eastern side to Pennsylvanian rocks on the west. The high peaks of the San Pedro Mountains, including the highest, Oro Quay Peak, are similar in composition to the central intrusion of the Ortiz Mountains. Rocks of the Ortiz porphyry belt commonly contain mineral deposits, including gold, as in the Ortiz Mountains. The scars on the hillsides ahead are copper and gold mines in skarn (lime-bearing silicate) deposits of Pennsylvanian limestones. Several mining companies have operated here, culminating in the 1970s with Gold Fields Corporation. Large garnet deposits still exist in the San Pedro mine. (0.3)

9.4 North-dipping Madera Group carbonates in roadcut. (0.5)

9.9 Intersection with NM–344 on right, which leads to San Pedro and (ultimately) to Edgewood and I–40. **Continue straight on NM–14.** (0.2)

10.1 MP 16. The Tijeras fault zone is now to our left. The ridge to our left contains Proterozoic rocks within the fault zone horst. The high peak to the right in the San Pedro Mountains is Cerro Columbo, elevation 7,572 feet, and consists of Tertiary intrusive rocks in Madera Group strata. (0.4)

10.5 Spectacular view of the Ortiz

Mountains ahead. The main mass of the Ortiz Mountains consists of monzonite that was intruded approximately 29 million years ago. The straight, gently west-dipping surface that forms a skirt around the Ortiz Mountains is known as the Ortiz surface. The surface dips away from the mountains in all directions and is capped by a thin layer of gravel known as the Tuerto gravel. Near the mountains, the Ortiz surface is eroded into the bedrock, whereas farther into the valley, it is cut into Santa Fe Group gravels. The surface represents an ancient erosional surface that sloped from the mountains to a former

The Ortiz Mountains from Highway 14.

The third and final mill at the Ortiz lode gold mine in 1901. Discovered in 1833, it was worked until the 1930s.

THE ORTIZ MINE GRANT

The Ortiz Mine Grant, in the heart of the oldest mining district in the country, played a prominent role in the development of New Mexico and the western U.S. The grant provided the funds that made the Santa Fe Trail trade economically significant and spurred U.S. interest in acquiring the Southwest. Colonists who founded New Mexico in 1598 came mostly from the silver mining town of Zacatecas, Mexico. By 1601 there were silver mines in the Cerrillos Hills, 10 miles north of the Ortiz Mountains. Though no record of Spanish mining in the Ortiz Mountains is known, it occurred all around the range. Two smelter sites southwest of the Ortiz were dated between 1650 and 1700, and in 1695 Governor Vargas founded Réal de los Cerrillos, a mining camp that served the Cerrillos silver mines. News of the Ortiz gold rush (the first gold rush in the western U.S.) was published in 1828, two decades before the California gold rush. In 1832 wealthy individuals began to purchase small lode mines in the Ortiz Mountains and to apply for official Mexican lode mining grants. In 1833 two local merchants, José Francisco Ortiz and Ignacio Cano bought the Santa Rosalia mine and applied for a mine grant. Their mine grant, later called the Ortiz Mine Grant, was the first grant in the area recognized by the U.S. government.

On February 1, 1858, the New Mexico legislature passed a law incorporating the New Mexico Mining Company as the first New Mexico corporation. The Ortiz Mine Grant of 69,458 acres was approved by Congress on March 1, 1861. After the Civil War, the New Mexico Mining Company raised money for mine development and construction of a larger stamp mill at Dolores, 4 miles southeast of Madrid. In 1868 New Mexico's first railroad was built. The 1.5-mile-long railroad connected the Ortiz mine to the new mill in Dolores.

In 1879 Steven B. Elkins and Jerome B. Chaffee gained control of the New Mexico Mining Company. In 1899 a 99-year lease for the entire grant, not previously leased to others, was sold to the Galisteo Company, owned by the Hoyt brothers of New York City. The Hoyts believed that large placer deposits still existed on the grant, and that Thomas Edison had developed a dry placer mill that would make large-scale placer mining profitable. However, Edison's engineers could not locate enough paying placer ground to justify building a large dry placer mill, and Edison abandoned the project. The Hoyts retained the Ortiz Mine Grant until 1943 when it was sold to a local livestock association, which in turn sold it to Mrs. George Potter in 1946. The Potters sold the surface but retained the mineral rights. Potter took in a partner and formed the Ortiz Mines Company of Missouri to manage the leasing of the mineral rights. Gold Fields Corporation obtained a mineral lease to 36,000 acres in the 1970s and started developing an open pit mine on the Cunningham deposit.

Through 1979 the vast majority of gold was produced by placer mining. Total production for the area before 1900 is estimated at 100,000 ounces, mainly produced during the Mexican period from placer deposits. Less than 2,000 ounces of gold was mined between 1900 and 1979. From 1979 to 1987, Gold Fields Corporation extracted approximately 250,000 ounces from the Cunningham deposit. LAC Minerals Corporation purchased the mineral lease in the late 1980s and brought in a potential partner, Pegasus Gold Corporation. Pegasus mapped two deposits on the south side of the Ortiz Mountains before abandoning the project in 1993.

Adapted from Homer Milford's article in the 1995 New Mexico Geological Society guidebook, Geology of the Santa Fe Region.

level of the Rio Grande that was nearly 1,000 feet higher than the present river level. (0.6)

11.1 MP 17. Entering Golden, site of the first gold rush west of the Mississippi in 1825. (0.3)

11.4 As the highway turns north into Golden we again cross the Tijeras fault zone. The Madera Group is exposed on the left side of the road and San Andres Formation ahead on the right. (0.7)

12.1 MP 18. Road bends left. Crossing northern splay of Tijeras fault zone. (0.3)

12.4 Crossing creek. Dark red mudstone and sandstone of Triassic Chinle Group exposed in left roadcut. Road straightens north and climbs up onto the Ortiz surface. (0.7)

13.1 MP 19. Crossing under powerlines. We are near the western edge of the Ortiz Mine Grant, which was deeded to Francisco Ortiz in 1832 by the Mexican government. Surface and mineral rights of the area were originally about 100 square miles, centered in the Ortiz Mountains. The surface and mineral rights were separated in 1948. The mineral rights continue to be held mostly by mining companies, and the surface rights by ranchers. (1.0)

14.1 MP 20. View of Sandia Mountains from 7:00–9:00, with the downhill ski runs visible near the crest at 8:00. To our right, this part of the Ortiz Mountains consists (from base to peak) of Triassic Chinle Group, Jurassic Todilto, Entrada, and Morrison Formations, and Cretaceous Dakota Sandstone and Mancos Shale. The Cretaceous strata are laced with Tertiary

The Cunningham Mine, Ortiz Mountains, early 1980s.

GOLDEN

Church records show that a small settlement existed in the San Pedro Mountains by 1776. Réal de San Francisco was one of two small mining camps settled in 1839 when placer gold was discovered on Tuerto Creek. (The other camp, Tuerto, has disappeared.) At first, the towns totaled only a few hundred Hispanic and Indian families who worked the mines by hand and transported the ore by burro, but by the mid-1800s up to 2,000 miners worked the gravels during the winter months. In the 1880s, when large mining companies arrived, the town's name was changed to Golden in anticipation of the riches to be derived from the mines. The change in name did not forecast reality, however. The veins soon pinched out, and water sufficient for sluicing gold from the gravels was unavailable. Fortunately for the towns, copper was discovered and kept the mines operating well into the 20th century. By 1960 more than 12 million pounds of copper had been produced in the New Placers district. Residents of Golden claim that heavy rains will expose nuggets of gold in the otherwise dry arroyos. Large reserves of lode gold do still exist in the nearby Ortiz Mountains.

intrusive rocks. Much of the high part of the mountain at 2:00 is composed entirely of these erosionally resistant Tertiary rocks. (0.7)

14.8 Panoramic view. At 9:00 the Colorado Plateau is visible in the far distance; the southern end of the Nacimiento Mountains merges northeastward with the higher and more rugged Jemez Mountains.

Los Lomas de la Bolsa laccolith at mile 16.3.

The valley of the Rio Grande and the Rio Grande rift are visible in the middle distance. The highest peak on the Jemez skyline is Redondo Peak, a 11,254 foot resurgent volcanic dome (uplifted after the caldera collapsed) within the Valles caldera. Cabezon Peak is visible on the western horizon. (1.5)

Cretaceous snails from the Mesaverde Group.

16.3 Road bends to right. The high knob ahead of us is part of the Los Lomas de la Bolsa laccolith, a 34-million-year-old intrusion in the Mancos Shale. At 10:00, near the base of the Jemez Mountains in the far distance, are bright white rocks of the Bandelier Tuff. (0.3)

16.6 STOP 1 – Turn left onto County Road 52A, then pull off and stop. This well-graded gravel road joins I–25 near the Santo Domingo Pueblo. The view from here provides a remarkable panorama to the west. Visible are the Sandias, Cabezon Peak, the Jemez Mountains, and the Rio Grande rift.

Return to NM-14 and continue north. Ahead, the road bends sharply right and enters Stagecoach Canyon. (0.5)

17.1 Mancos Shale is exposed in roadcut on left, and ahead. (1.0)

18.1 MP 24. Tertiary intrusive rocks are exposed ahead in left roadcut at big turn. (0.8)

18.9 We are driving on the Ortiz surface. Gravel pit to the left is excavated in gravel deposits at top of surface. (0.8)

19.7 Tertiary intrusive units are exposed on ridges to both right and left of road. (1.4)

21.1 MP 27. The canyon and highway are on Cretaceous Mesaverde Group sedimentary rocks. Both sides of the valley are capped by Tertiary intrusive rocks. Notice the black coal spoil piles to the right. Coal found in small assay smelters suggests that coal may have been mined in this area before 1800 for assaying or smelting. Spanish miners were mining anthracite at Madrid by 1835, but mining began in a big way when the AT&SF Railroad reached the area in 1880. Up to 45,000 tons of coal was mined annually between 1888 and the 1950s, when the demand for coal dwindled

TURQUOISE

The word turquoise is derived from the French word Turquie for Turkey, where the mineral was thought to have originally been mined. Turquoise is a hydrous aluminum phosphate that contains a small amount of copper and occasionally a bit of iron. The blue color of the gem is due to copper, whereas the green color is due to iron. Uniformly colored blue specimens are greatly in demand for fine jewelry, often matched with gold, silver, and diamond. Turquoise can range in color from a hard, bright, electric blue to pale blue to pale sea-green. Crystals are very rare and greatly valued. In the southwestern U.S. an unusual occurrence, formed by a network of thin chestnut-brown veinlets in the turquoise, is greatly admired. This "spider" variety is notable at Bisbee and Morenci, Arizona.

The recent interest in turquoise jewelry made by southwestern Native American Indians is only the latest in a long history of appreciation for this striking mineral. Turquoise was mined on the Sinai Peninsula 7,000 years ago. After being long abandoned, these rich mines were rediscovered in 1849. Egyptian artists fashioned blue turquoise into grand broaches and pectorals and created fine ornamental inlays.

Neither the Greeks nor the Romans used turquoise, and it wasn't until the Middle Ages that the fine blue turquoise of Persia was mined. India, Burma, and Ceylon supplied no turquoise, although Tibetans used Chinese turquoise widely. During several Chinese dynasties, from 1766 B.C. to 221 A.D., Chinese turquoise was crafted into stunning artistic pieces worn by royalty. In spite of the enormous copper deposits of South America, surprisingly little turquoise was used there.

Tiffany Mine.

Turquoise for the inlaid ritual objects of the Aztecs was mined in New Mexico and Arizona by local Indian tribes and traded south. Evidence for this early trading has been found by neutron activation techniques that "fingerprint" Aztec turquoise with North American sources. The major trade came from Mt. Chalchihuitl in the Cerrillos area; it was extensive and contributed to a lavish trade in gem, ceremonial, and religious art. Aztec artifacts from Alta Vista, Mexico (northwest of Mexico City), a prehistoric site dated at A.D. 700, contain turquoise from Cerrillos. By inference, turquoise mining at Chalchihuitl extends back nearly 13 centuries. The earliest use of turquoise as gem has been documented in a burial site near Mezcala, Guerrero State, Mexico, dated archaeologically at 600 B.C. If this turquoise came from the Cerrillos area, as is possible, then that mining date is pushed back to nearly 26 centuries, making it the oldest mining in North America. Mount Chalchihuitl, here in the Cerrillos Hills, was the largest open pit mine of any metal or mineral in North America when the Spanish arrived in 1535.

Prehistoric drilled turquoise from Chaco Canyon.

Madrid, ca. 1935.

and the mines were closed. (0.7)

21.8 Red hills on right are baked and altered (through contact metamorphism) sedimentary rocks of the Mesaverde Group. Notice sculptures on hills. (0.6)

22.4 Entering Madrid. The town of Madrid may have been named for Madrid, Spain, but it more likely takes its name from an early Spanish family who arrived in New Mexico in 1603. One of Diego de Vargas's captains, Roque Madrid, had interests in lead mines in the area, and members of the family may have been living here when coal mining began in 1835. After 1880 Madrid became a company town. It became a ghost town for 20 years after the coal mines closed in 1954. Today Madrid is a mecca for artists and tourists. (0.3)

22.7 The Old Coal Mine Museum, Engine House Theater, and Mine Shaft Tavern are well worth a visit. The museum contains a coal seam in the Mesaverde Group and a 1900 vintage steam locomotive. Many walls and walks in Madrid are constructed from local Mesaverde Group sandstone. (1.1)

23.8 Good exposures of the Mesaverde Group may be found in the drainage along the road on the right. In this area the Ortiz surface is eroded on Cretaceous bedrock and overlain by the Tuerto gravel. The flanks of the Ortiz Mountains have been extensively explored for placer gold, mainly in the 1800s. Most of the 100,000 ounces of placer gold production in the Ortiz Mountains came from the Cunningham Hill area several miles east of here. A limiting factor in separating the precious metals from the gravels has been the lack of water. The most notable miner was Thomas Edison, who invented an electro-winnowing plant for working dry placers, which is said to have failed miserably. (0.3)

24.1 MP 30. Straight ahead, beyond the notch in the far skyline is the southern end of the Sangre de Cristo Mountains in the Santa Fe area. At 11:00, in the middle

CERRILLOS

The Cerrillos mining district is renowned worldwide for its turquoise mines. Many prehistoric and Spanish Colonial mines are scattered over the Cerrillos district. Most are small, but one of the larger mines north of Cerrillos illustrates what indefatigable effort with hand tools can accomplish. The turquoise mine at Mt. Chalchihuitl is the most extensive prehistoric mining operation in America. Worked for nearly 1,000 years with stone tools, the pit was cut to a depth of 200 feet through solid stone. Chalchihuitl is from the Nahuatl (Aztec) word for turquoise, xiuitl, and was apparently named by Indians from the Valley of Mexico who accompanied the Spanish to New Mexico.

The village of Cerrillos got its start in 1879 when silver, gold, and copper were rediscovered in the adjacent Cerrillos Hills. When its population peaked in the 1880s, the town contained 21 saloons and four hotels. In the 1880s and '90s, Ulysses S. Grant and Thomas Edison were both guests at the luxurious Palace Hotel. Edison was in Cerrillos to conduct experiments in extracting gold from sand and gravel by static electricity. Mining activity greatly decreased by the end of the 1800s. The rustic store fronts and dirt streets have served as the backdrop for many Hollywood movies.

Besides mines and mining towns, the Galisteo Basin is famous as the location of several large and medium-sized pueblos, many of which were inhabited when the Pueblo Rebellion occurred in 1680. Residents of the Galisteo pueblos, suspected of being major players in inciting and executing the rebellion, abandoned their towns and, except for one desultory trial, refused to reinhabit them when the Spanish returned in 1692 and attempted to resettle Pueblo people in their old homes. (Hano, a Tewa-speaking village on the Hopi mesas in Arizona, was settled in 1700 by people from the Galisteo Basin, and they are still there, 300 years later.) Nearly all of the sites (San Lázaro, Galisteo Pueblo, San Cristóbal, Pueblo Largo, San Marcos, Pueblo Blanco, Pueblo Colorado, and Pueblo Shé) are on private land, but the Archaeological Conservancy owns most of San Marcos. The 2,000-room pueblo and 17th century mission is fenced, but may be visited with special permission from the conservancy.

Galisteo Creek, Cerrillos, ca. 1895-1900.

distance, the high hills are part of the Cerrillos Hills. The best known mine, the Chalchihuitl mine, is located in Tertiary intrusive rock near Mt. Chalchihuitl (6,976 feet), the highest peak in the range. (1.6)

25.7 Crossing bridge over Galisteo Creek.(0.1)

25.8 Turn left to Cerrillos. End of this leg.

Devil's Throne, Stop 2.

0.0 Reset odometer to 0.0 at intersection of NM–14 and Main Street and drive west on Main Street into Cerrillos. This log begins at the intersection of NM–14 and Main Street in Cerrillos and ends at I–25, at the top of the La Bajada fault escarpment. Most of this route is on a well-graded gravel road, generally passable for two-wheel-drive vehicles in dry weather. The road bed, however, is constructed on clay-rich sediments and should not be attempted by any type of vehicle when the road is wet. (0.4)

0.4 Stop sign. Intersection of Main Street and First Street in the center of the village of Cerrillos. Turn right onto First Street (0.1)

0.5 Cross Santa Fe (now Burlington Northern Santa Fe) Railroad tracks. Pavement ends. Turn left immediately onto Waldo Canyon Road (County Road 57), paralleling railroad tracks. (0.1)

0.6 Cross San Marcos Arroyo. Cattle guard on east side. (0.2)

0.8 View to the south, across Galisteo Creek, of the Ortiz Mountains. (0.1)

0.9 View straight ahead of Devil's Throne. (0.2)

1.1 STOP 2 - Devil's Throne. Pull off on left side of road. On the south side of the road is the feature known as Devil's Throne, an irregularly shaped intrusion of andesite porphyry. On the north side of the road, just east of Devil's Throne, the outcrop exposes the contact between this intrusion and the Cretaceous Mancos Shale, into which it has been intruded. The intrusion itself is about 30 million years old. This intrusion is responsible for much of the deformation we see in the adjacent sedimentary strata here. (0.2)

1.3 Road climbs steep grade in a narrow belt of Mancos Shale flanked on both north and south by intrusive rock. Contact of contact-metamorphosed shale and Devil's Throne intrusive is exposed on left side of road. (0.3)

1.6 Mancos Shale on left contains an intrusive sill (an intrusion that was emplaced parallel to the bedding of the shale) that displays columnar joints. Columnar jointing results from contraction that occurs during cooling of the intrusion; the polygonal shrinkage cracks give rise to these distinctive parallel prismatic columns. (0.6)

2.2 The cottonwood grove on the left marks the site of the town of Waldo. This small community was founded in the early 1880s and flourished in 1892, when it became the junction for the Santa Fe Railroad and a spur line to coal mines in Madrid, 12 miles to the south. At one time coal was also mined in Waldo Gulch, the most significant tributary to Galisteo Creek, which is visible almost directly to the south. For about 10 years this coal was roasted in coke ovens here in Waldo. After the local mines were closed, Waldo continued to serve in a support role to the continually growing coal-mining community at Madrid. Most important, all of Madrid's water (150,000 gallons per day) was hauled in large tank cars, four times daily, from wells at Waldo. When the Madrid coal mines closed in 1954, Waldo quickly turned into a ghost town. Little remains today. (0.6)

2.8 Roadcut contains a north-trending dike that has intruded Mancos Shale. More than a dozen such dikes radiate like wheel spokes from a point in the northern Cerrillos Hills. (0.9)

3.7 Roadcut on right through another dike cutting Mancos Shale. Good view to right of the Cerrillos Hills. The highest point is Cerro Bonanza (7,088 feet). (0.2)

Outcrop on the north side of the road at Stop 2, just east of Devil's Throne.

3.9 Roadcut exposes a ridge-forming dike that has baked the country rock shale. (0.5)

4.4 Roadcut in Mancos Shale (0.7)

5.1 Ridge crossing from right to left beyond curve in road is held up by a dike intruded into the Mancos Shale. This dike is exposed at Stop 3 ahead. (0.3)

5.4 STOP 3- Pull off onto right shoulder and park. Views to the south include the Ortiz Mountains flanked by the Ortiz surface. This surface, which is approximately 3 million years old, is generally regarded as the highest and the oldest surface in the Rio Grande valley of the Albuquerque area. Here the surface is formed on top of the Tuerto gravel (which ranges in age from 2 million years to the present), which rests on an older erosion surface. The source for these gravels is the Ortiz Mountains. To the right of the eroded mar-

Mancos Shale

gin of the Ortiz surface are east-dipping cuestas of Cretaceous Menefee Formation sandstones and sills intruded into the Upper Cretaceous rocks. These eastward dips characterize the local structure in the footwall of La Bajada fault, which trends northward from near the west base of the Ortiz Mountains. La Bajada fault marks the eastern margin of the Santo Domingo Basin within the Rio Grande rift. It is the northernmost of a series of right-stepping normal faults that form the eastern margin of the Rio Grande rift north of the Sandia Mountains.

Roadcuts here expose the Niobrara Member of the Mancos Shale. Concretions as large as 6 feet in diameter have weathered out of the shale to form brown lumps on adjacent hillsides. The Mancos Shale here is a fissile (easily split along closely spaced bedding planes) siltstone containing many trace fossils, of which Chondrites (fossilized tunnel structures, dwelling or feeding burrows made by a marine worm) is most abundant. Thin interbedded very fine grained sandstone layers contain crossbedding and abundant fine terrestrial plant fragments that are conspicuously absent in the dark siltstone. The sandstone layers are interpreted as storm beds. Shells of bivalves (oysters and fragments of large inoceramid shells), molds and casts of high-spired gastropods, and rare ammonoids (coiled and straight varieties) are present as fossils in this roadcut and on the adjacent hillslopes. A 15-foot-wide dike cuts the shale at the west end of the roadcut.

The road crosses private property on either side of the road in this vicinity; please respect the wishes of the landowner and do not venture onto private land. Continue driving westward. (0.3)

5.7 View to west (left) of Cuerbio basalt (Pliocene age) resting unconformably on Mesozoic sedimentary rocks next to La Bajada fault scarp. Southern part of the Jemez Mountains volcanic field forms the western skyline on the west side of the rift. The Cerrillos Hills are visible to the east. (0.4)

6.1 Dike exposed along wash on right. This basaltic dike is one of many such dikes in this area that trend north–south, parallel to La Bajada fault. These dikes are unrelated to Cerrillos Hills magmatism but may be related to the Pliocene basalts of Cerros del Rio volcanic field to the north. (0.5)

6.6 Alluvial and eolian (windblown) sediments of the Tuerto gravel rest on an erosional surface on Mancos Shale. (0.2)

6.8 Cattle guard; Cuerbio basalt visible along streamcuts to left and right of the road. This basalt is 2.7 million years old and forms the Mesita de Juana Lopez, the southern edge of the Cerros del Rio volcanic field. (0.2)

7.0 View straight ahead of Tetilla Peak, a dacite dome atop an andesite shield within the Cerros del Rio volcanic field. An unnamed cinder cone can be seen by I–25. Jemez Mountains form the western skyline. Southern Sangre de Cristo Mountains visible to the north. (0.3)

7.3 Pavement resumes. (1.0)

OPPOSITE: Aerial view of I-25 descending the La Bajada escarpment.

8.3 STOP 4. Pull off road at intersection with frontage road. Once on I-25 south, we will descend the escarpment of La Bajada. Two major branches of the Camino Real were established during the Spanish Colonial period to carry traffic over this escarpment, the highest and most difficult barrier between Chihuahua and Santa Fe. The earliest road, built in 1609 or 1610 when Santa Fe was founded, followed the Santa Fe River Canyon through the escarpment, but high water often made the road impassable. A second route, avoiding the river by traversing La Bajada Hill parallel to but above the river, was established in the mid-1700s. The steep face and erosion were constant problems. John Hansen Beadle, who traveled over La Bajada road in 1872, wrote: "Down the face of this frightful hill the road winds in a series of zig zags, bounded in the worst places by rocky walls, descending fifteen hundred feet in three-quarters of a mile." A third branch, apparently established during the Mexican period and improved by the U.S. Military during the 1846 invasion of New Mexico, avoided both the river gorge and the escarpment. Instead, it crossed the Mesita de Juana Lopez and continued south across Galisteo Creek to the pueblos of Santo Domingo and San Felipe. This branch was used until the late 1800s and early 1900s. The route of I-25 over the escarpment was established in the late 1970s. (0.1)

8.4 Intersection with I-25; end of County Road 57. This log continues southward to Albuquerque on I-25.

0.0 Reset trip meter to 0.0 and turn south on I-25 on-ramp. (Exit 267). (0.3)

0.3 Merging on southbound I-25. Jemez Mountains are visible on the skyline at 3:00. We are traveling on Mesita de Juana Lopez, an alluvium-covered surface underlain by basalt flows. (0.7)

1.0 Begin descent over La Bajada escarpment. Note columnar jointing

> *Down the face of this frightful hill the road winds in a series of zig zags, bounded in the worst places by rocky walls, descending fifteen hundred feet in three-quarters of a mile.*
>
> — JOHN HANSEN BEADLE, 1872

in basalt flow capping the surface. The 2.8 million-year-old basalt rests on a thin layer of upper Santa Fe Group deposits that overlie gray shale and sandstone of the Mancos Shale. (0.4)

1.4 At 9:00, contact between pinkish sediments of the Santa Fe Group and underlying Mancos Shale. Note the dark basaltic dike intruding the Mancos Shale. (0.4)

1.8 Light-brown sandstone of lower Mancos Shale in roadcut on left. Note mantle of colluvium with basalt blocks on steep slopes. Ahead on left is a dark-red igneous dike cutting Morrison Formation sandstone. Just ahead, on the west end of the roadcut, note the fault contact between the yellowish Morrison Formation and gray Mancos Shale. Be on the lookout for more faults as we descend the hill. (0.2)

This fossil palm log from the Galisteo Formation indicates a warm, subtropical Eocene climate.

2.2 Exit 264 to Cochiti Pueblo. Roadcuts in Galisteo Formation here are unconformably overlain by Espinaso Formation. Both the Eocene Galisteo and Oligocene Espinaso Formations are seen only within this part of the scenic trip area. The brown to reddish sandstone and shale of the Galisteo Formation represent deposition in a floodplain environment. The Espinaso Formation contains mainly detritus from eroding volcanoes to the northeast, interbedded with ash-flow tuff (layers of volcanic rock deposited by a flow of volcanic ash and gases) layers. To the east, this sequence is faulted against Mancos Shale.

Cochiti Pueblo was established at its present location before the arrival of the Spanish in 1540. The northernmost of the Keresan-speaking pueblos, Cochiti is famous for cylindrical drums made of hollowed-out cottonwood logs, carefully maintained tradition in ceremonials and costuming, and fine pottery bowls, jars, animals, and storyteller figures. (0.3)

2.5 Crossing La Bajada (Rosario) fault; light-pink Santa Fe Group beds on the west side are faulted against the Galisteo Formation. This fault generally marks the boundary between the Española Basin to the north and the Santo Domingo Basin to the south. Both basins are part of the Rio Grande rift. (0.5)

3.0 Entering Santo Domingo Pueblo. Begin descent into Galisteo River drainage. Santo Domingo Pueblo is one of the best-known and largest tribes in the Southwest. Several feast days are held during the year. The Corn Dance, August 4th, is especially spectacular with its many beautifully costumed dancers. Santo Domingo men have been famous as traders for many years, traveling as far as California and Oklahoma to trade in Santo Domingo pottery, jewelry, moccasins, woven belts, and goods produced by other tribes. (0.3)

3.3 Reddish-brown alluvium derived from

Galisteo Formation red beds exposed in La Bajada escarpment covers downfaulted upper Santa Fe Group deposits. (0.7)

4.0 Crossing main line of Burlington Northern Santa Fe Railroad (formerly AT&SF). Roadcuts ahead in Santa Fe Group conglomeratic sandstone are overlain by a narrow strip of inset terrace (a stream terrace formed during successive periods of vertical and lateral erosion such that remnants of a former valley floor are preserved on the valley walls) deposits along Galisteo Creek. (0.3)

4.3 Crossing Galisteo Creek. To the east you can see a quarry on the north side of the creek, where gypsum from the Todilto Formation is mined.

Galisteo Dam and Reservoir, located farther upstream, prevent flash floods from racing down the creek. The dam was authorized by the 1960 Flood Control Act for flood control and sediment retention but does not impound a permanent lake. Access to the dam is 4.6 miles eastward from I–25 to the south end of the dam embankment and beyond a locked gate. Vehicle entrance is restricted, but a visitor center and picnic area are nearby on an unrestricted spur road to the right of the locked gate. (1.2)

GALISTEO DAM

The dam (at far left in photo) is a rolled earth-fill structure 2,820 feet long with a maximum height of 158 feet above stream bed, a crest width of 20 feet, base width of 2,600 feet, and a compacted fill volume of 9.33 million cubic yards. The outlet works, located on a bedrock-excavated bench in the right abutment, includes an inlet channel, an ungated intake structure (one of only two such structures in the state operated by the U.S. Army Corps of Engineers, the other being Santa Rosa Dam, completed in 1980), a 10-foot-diameter conduit, a flip bucket (for energy dissipation), and an outlet channel. The ungated outlet works allows Galisteo Creek to flow naturally during times of low rainfall but provides a uniform discharge during high rainfall or flash flood events.

Elevation of the dam embankment is 5,639 feet, allowing for a maximum pool elevation of 5,634 feet above sea level. At maximum pool height, the reservoir has a surface area of 2,920 acres with a capacity of 153,400 acre-feet and storage allocations for 10,200 acre-feet of sediment below the spillway crest. The spillway (elev. 5,608 feet) has a capacity of 90,000 cubic feet per second at maximum pool height. Drainage area behind the dam is 596 square miles. Construction on the dam began in March 1967 and ended in September 1970. The maximum pool of record was July 22, 1971, when a pool elevation of 5,517 feet was recorded, corresponding to 2,870 acre-feet of water.

5.5 MP 262. Route ahead crosses several surface remnants underlain by thin alluvial veneers and upper Santa Fe Group deposits. The surfaces represent the top of the Santa Fe Group. Juniper with scattered piñon are present along both sides of the freeway. (3.0)

8.5 MP 259. For the next 2 miles, we will be traveling on gravels that cover the upper Santa Fe Group, which is exposed in arroyos. (1.0)

9.5 Exit 257 to Budaghers; leaving Santo Domingo Pueblo. This settlement is named for Saith Budagher who arrived here in the late 1800s from Lebanon and established several businesses. His descendants are still Sandoval County residents. (1.0)

10.5 MP 257. Espinaso Ridge at 9:00. This ridge is composed of coarse volcanic breccia and fanglomerate (a conglomerate composed of slightly waterworn fragments of all sizes deposited in an alluvial fan) of the Espinaso Formation. Along the western base of the ridge, the eastward-dipping Espinaso Formation conformably overlies the Galisteo Formation. (0.9)

11.4 Entering San Felipe Reservation. (1.1)

12.5 MP 255. Begin descending valley slope cut in upper Santa Fe Group

SAN FELIPE

Like Cochiti, Santo Domingo, Zia, and Santa Ana, San Felipe is an eastern Keresan-speaking pueblo. Legend says that Cochiti and San Felipe were one people and lived with other Keresan speakers at Frijoles Canyon (in Bandelier National Monument) and other places until they were guided to their present locations by the Corn Mother Spirit. When the Spanish arrived, the San Felipe people were living in twin villages facing each other across the Rio Grande.

The dances at San Felipe are considered by many to be among the finest of the Rio Grande pueblos.

deposits derived from various rocks to the east. (1.4)

13.9 The ancestral river gravels in the roadcuts ahead can be traced along the Rio Grande valley from here to El Paso, Texas. These upper Santa Fe Group deposits represent a time when the Rio Grande was a broader river of braided channels choked with sand and gravel. This is quite different from the narrower, meandering river of today. (0.3)

14.2 Volcanic ash bed with some pumice clasts exposed in roadcut ahead. This ash was derived from a Jemez volcano eruption that occurred about 1.5 million years ago. (0.5)

14.7 Crossing broad Tonque Arroyo near Exit 252 and access to San Felipe Pueblo. This arroyo drains parts of the Sandia, Ortiz, and Monte Largo highlands. Visible on the south side of Tonque Arroyo to the left of the San Felipe casino is the "Big Cut" that was constructed before 1926 as part of the original course of Route 66. (2.6)

17.3 Descending onto an alluvial apron formed by coalescing alluvial fans. These deposits probably bury an inset terrace of the Rio Grande. Across the river at 3:00 is Santa Ana Mesa, capped by San Felipe basalt dated at 2.5 million years. Exposed beneath the basalt are upper Santa Fe Group ancestral-river sediments (light gray) overlying piedmont sediments (reddish brown). Note that the basalt flow is offset by a series of faults, resulting in a mesa that steps down to the south. This fault system separates the Santo Domingo Basin from the Albuquerque Basin. (2.9)

20.2 Algodones exit. The town of Algodones is located along the river to the right. The town name may derive from the "cotton" produced by abundant cottonwood trees. To the west, along the valley margin, are inset terrace gravels. Quarries along here are extracting gravels from these terrace deposits and the axial river sediments for use as aggregate and in concrete. (1.1)

21.3 Entering Santa Ana Pueblo. Good view of Sandia Mountains and Rincon Ridge at 10:00. We are traveling on an abandoned surface of Las Huertas Creek fan. Downcutting by the creek has resulted in the surface being cut off from more sedimentation. This surface projects to a base level 120 feet above the present Rio Grande. Just south of here, we cross the Algodones fault zone and enter the Albuquerque Basin. Large sand and gravel operation on left.

Santa Ana Pueblo consists of two villages. The old village, Támáyá, 8 miles west of the Rio Grande on the north bank of the Jemez River was settled by the late 1500s. However, good agricultural land was limited, and the pueblo started acquiring land along the Rio Grande in the 1700s. People began by spending a few days and then the entire growing season near the fields. Ultimately, the new village of Ranchitos (visible from I–25) was populated year-round. The old village was maintained as a ceremonial village. Recently, some who seek a quieter more traditional life have returned to the old village. Those who prefer modern conveniences continue to live at Ranchitos. (1.2)

22.5 MP 245. Canjilon Hill is at 2:00. This hill forms a small, oval mesa about 4,000 feet long by 2,000 feet wide. It is a special type of volcano called a maar, created by steam explosions when molten rock encounters a water-saturated zone and the water flashes to steam. The resulting explosions produce a low crater. Several dikes intruded the hill, and a lava lake filled its southern portion. Canjilon Hill is similar in age to the San Felipe basalt, about 2.5 million years old. Roadcuts expose more inset terrace gravels that overlie Santa Fe Group sediments. (1.3)

Canjilon Hill.

23.8 Large gypsum processing plant operated by Centex American Gypsum Company is on the right. Gypsum is mined from the Todilto Formation near San Ysidro about 20 miles to the west. Trucks haul the gypsum here, where it is processed into various kinds of construction-grade wallboard. (1.2)

25.0 Exit 242 to Bernalillo. Outcrop on left exposes middle Santa Fe Group deposits with an inset terrace gravel.

SCENIC TRIP FOUR - (46.3 MILES)
Along the Rio Grande

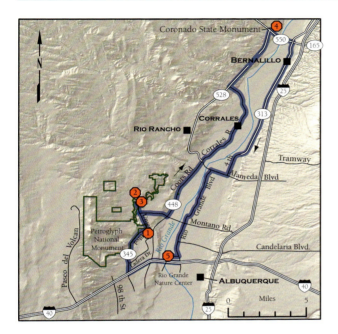

This trip explores the river corridor and the rapidly developing West Mesa region of Albuquerque, including Corrales, Rio Rancho, and Bernalillo. The trip first focuses on the lava flows and underlying deposits of the Volcano Cliffs, where abundant prehistoric rock etchings at Petroglyph National Monument provide opportunities to learn about prehistoric Indian art. The rest of the trip parallels the Rio Grande, examining the geological development of this great river and the elaborate irrigation system that serves the region. The Albuquerque volcanoes can be seen from certain vista points along the cliffs, but you cannot drive across the monument to them. Those interested in a closer look may visit these unique volcanoes on Trip 5.

From Petroglyph National Monument, the trip continues northward along the Rio Grande through the charming and historic community of Corrales and the modern development of Rio Rancho. At the intersection with NM–550 (shown on older maps as NM-44), the trip heads east, with a stop at Coronado State Monument. The painted kiva murals on display at Coronado State Monument are among the highlights of the trip. We then turn south on NM–313 through Bernalillo and Sandia Pueblo. We will discuss the historical significance of these pueblos and early Spanish settlements. The trip ends at the Rio Grande Nature Center. We recommend a full day for this trip, to allow time for recommended stops and spontaneous detours.

0.0 Begin on I–40 at Rio Grande Boulevard overpass. Passing over Rio Grande Boulevard. Continue west on I-40. We are now traveling on the Rio Grande floodplain on which central Albuquerque was built. This area along the river is prone to flooding because it is at a lower elevation than the river channel. Although a dirt levee system normally keeps the river contained, it forces the river to aggrade (build up sediments within the channel), thus increasing the potential for severe flooding by increasing the difference

THE RIO GRANDE BOSQUE

One of the most endangered ecosystems in the United States is the cottonwood–willow bosque, a unique combination of wetland and forest. In central New Mexico the bosque consists of a narrow ribbon of green, nearly continuous for over 150 miles along the Rio Grande, from Cochiti Reservoir near Santa Fe to Elephant Butte Reservoir at Truth or Consequences.

Sixty years ago, much of the riverbank near Albuquerque was too wet and alkaline to support cottonwoods and willows. Most of this waterlogging was a direct result of human settlement activities including agriculture and logging along the river and its northern tributaries. Clearing of the land, overgrazing, and forest clearing from Colorado to central New Mexico all worked to increase the sediment load washed into the river. Small dams and diversions slowed the river, resulting in more aggradation (deposition) along the river bed. Artificial levees were constructed to keep the river from flooding into the floodplain, which increased aggradation. In the middle Rio Grande basin, as the river bed deposited silt, its channel was raised higher and higher above the valley floor.

Ground-water levels rose along with the river, and percolation from irrigation ditches raised the water levels even more. By 1920 the average depth to the water table in the central valley was 2.5 feet. Orchards were killed, and alfalfa fields were abandoned. Unhealthy swamps formed in Albuquerque, and homes and businesses sustained water damage. It was not until the 1930s, when the Middle Rio Grande Conservancy District drained the treeless marshland along the river, that the bosque became reestablished as a continuous ecosystem.

Interestingly, in the 1950s the bosque was viewed by some as a wasteful consumer of precious water, and proposals were made to have it clear cut. Fortunately, more rational opinions prevailed. Recently, however, questions have arisen about the ecological health of the river system and bosque. Non-native plant species such as salt cedar and Russian olive have begun to replace an aging and diseased population of native cottonwoods and willows. Water diversions designed to reduce flooding and deliver irrigation have greatly changed the natural river cycle. Declining water quality, due primarily to municipal, industrial, and agricultural discharges, as well as frequent dewatering and channelization of the river, has greatly changed the river ecology.

Historically the Rio Grande bosque was a commons for the people of the valley. It provided residents with water, food, wood for fuel and construction, and solitude. Early in this century the entire stretch of riparian habitat was deeded in trust to the Middle Rio Grande Conservancy District. The conservancy, along with local organizations and governments, protects this natural treasure from local threats such as fires, trash dumping, poaching, off-road vehicle use, and illegal wood cutting. Today the bosque is one of the Southwest's richest riparian areas.

in elevation between the river and the floodplain. This potential has been lessened somewhat by the construction of dams upstream. (1.2)

1.2 Crossing the Rio Grande and ascending out of the inner valley onto the Segundo Alto surface, an inset terrace of the Rio Grande. Note levees on each side of the river. The wooded area along the river is known as the bosque. Roadcuts expose sand and clayey silt of terrace deposits that underlie the surface. (0.6)

1.8 Passing under Coors Road. Continue west on I–40. We are traveling on the Pleistocene Segundo Alto surface situated about 140 feet above the Rio Grande. This surface represents a former valley floor before the Rio Grande cut the inner valley. (1.2)

3.0 Exit 154; exit onto Unser Boulevard (Highway 345). Stay in the far right lane. (0.9)

3.9 Crossing Ladera Drive. This area west of the river has experienced tremendous urban growth in the past 10 years and continues to be one of the fastest growing regions of New Mexico. (0.4)

4.3 Crossing 98th Street. Volcano Cliffs straight ahead. These cliffs are part of a 17-mile-long escarpment of basalt that overlies light-colored sand with interbeds of pink silt, gray clay, and gravel. The 10- to 20-foot-thick basalt flowed from the Albuquerque volcanoes about 2 miles to the west 100,000 years ago. The underlying sediments probably represent an inset terrace deposit that is younger than the Santa Fe Group. Basalt is more resistant to erosion than the valley-fill sediments so it tends to form prominent cliffs. (1.2)

5.5 Rinconada flood control channel. These channels are designed to safeguard the area from flooding. We are driving on a Holocene alluvial apron that extends from the base of Volcano Cliffs. (0.1)

5.6 Turnoff to Rinconada Canyon Kiosk; continue straight ahead. The area provides parking and information. A hiking trail around the canyon starts here. Picnics and wheeled vehicles are not permitted. (0.8)

6.4 Intersection with Western Trail. Turn left and follow signs to Las Imágenes Visitor Center. Note that the gates on this road are open only from 8 a.m. to 5 p.m. (0.4)

6.8 STOP 1 - Las Imagines Visitor Center, Petroglyph National Monument. This park was established in 1990 to protect the more than 15,000 petroglyphs scattered along the 17-mile-long basalt escarpment. It is jointly owned and administered by the City of Albuquerque and the National Park Service. Coated with desert varnish, a dark-colored, metallic-like veneer built up over the millennia by weathering processes, basalt provided an ideal canvas for prehistoric people. Considered sacred by Native Americans, the petroglyphs are carved and pecked into the rock surface, rather than painted onto the rock (pictographs). The motifs represented (kachina masks, birds, animals, horned serpents, lizards, humans, star being, and hunch-backed flute players) indicate that most were carved during the Pueblo IV period, A.D. 1300–1600, but a few (circles and meandering lines) were apparently carved between 1000 B.C. and A.D. 500; still others (crosses) were made by Spanish settlers after 1598. Park rangers can direct visitors to the best petroglyph viewing areas. The small bookstore and information center here is worth a visit; be sure to ask for a map of the park.

Retrace route back to Unser Boulevard and reset mileage at intersection.

0.0 Turn left onto Unser Boulevard (0.7)

0.7 Crossing San Antonio Arroyo. This drainage has been modified to hold flood runoff. (0.7)

1.4 Montaño Road intersection. Continue straight on Unser Boulevard. (0.6)

2.0 Entrance to Boca Negra Canyon, Petroglyph National Monument on right. **Continue on road up Boca Negra Canyon and onto mesa for overview of area before visiting monument. (0.3)**

2.3 **STOP 2. Turn right into parking area.** This stop provides an opportunity to examine the basalt flow and view the Albuquerque volcanic field. We are about 2 miles from the volcanoes.

Walk along the trail to the picnic shelters near the top of the cliff to the north to view the city and examine the basalt. From the picnic area, the Sandia Mountains form Albuquerque's backdrop. South of the Sandia Mountains are the Manzanita and Manzano Mountains. These eastward-tilted mountains form the prominent eastern boundary of the basin; they contain Precambrian granitic and metamorphic rocks overlain by layers of Pennsylvanian limestone and shale. From here we have a good view of the modern Rio Grande and its inner valley. We are about 15 miles from the mountain base. Vegetation here includes bunch grass, yucca, prickly pear cactus, and snakeweed.

The basalt flow here is covered by a thin layer of windblown sand and alluvium. Take a close look at the basalt: it is dense and contains large crystals of white plagioclase and green olivine. Note the vesicles in the rock, produced by gas bubbles that escaped as the lava flowed and cooled. These basalt flows may have moved across the mesa at 5–10 miles per hour.

To the west the Albuquerque volcanic field includes about 10 different basalt flows that erupted from at least ten small volcanoes and spatter cones. These volcanoes are aligned along a fault zone known as the County Dump fault, which served as a conduit for the molten rock to reach the surface. The flows erupted approximately 155,000 years ago. The most prominent cone, marked with a white J, is about 200 feet high with a diameter of 2,100 feet.

Archaeological excavations in the 1960s at Boca Negra Cave, a small air pocket in one of the northern volcanic cones, recovered cobs of Maíz de Ocho, an ancient species of corn. The cobs were dated by radiocarbon to A.D. 370. **Retrace route back along Unser Boulevard and turn left into the national monument.**

Vesicular basalt. White crystals are plagioclase.

STOP 3 - Petroglyph National Monument (fee area). Well-marked hiking trails lead to basalt boulders covered with various types of petroglyphs. As Albuquerque developed westward across the West Mesa, the population placed a greater and greater burden on the marvelous rock art along the basalt escarpment. Vandals desecrated and destroyed the delicate petroglyphs, and housing developments along the cliffs would have made the art off-limits to visitors. In response, citizens worked to establish Indian Petroglyph State Park and Volcano City Park. Finally in June 1990 the cliffs and volcanoes were protected with the creation of Petroglyph National Monument, which includes 17 miles of petroglyph-covered cliffs and the five extinct volcanoes. To resume trip, retrace route and turn left onto Montaño Road (mile 1.4). Mileage will reset at 0.0 at the intersection of Unser Boulevard and Montaño Road.

0.0 Drive east on Montaño Road. We are descending a Holocene alluvial apron that has formed from the base of the Volcano Cliffs scarp. The homes in this area have been built since 1990. (1.0)

1.0 Mariposa diversion channel was constructed to protect the homes from flash floods. (0.3)

1.3 Crossing Taylor Ranch Road. Merge into left lane. (0.2)

1.5 Descending from Segundo Alto surface

Eroded basalt boulders at the edge of West Mesa are covered with petroglyphs for which the monument is named.

THE WEST MESA

Given the natural boundaries and Pueblo Indian lands to the north, east, and south of Albuquerque, it's no surprise that the West Mesa is the city's last frontier. Much of the west side is being converted from grazing land and open space to a single, sprawling residential and industrial complex. The state's largest shopping mall, the 95-acre Cottonwood Mall, opened for business in 1996. Several factors have encouraged the west side development: the large tracts of land are easily developed, relatively inexpensive, and provide incomparable views and a sense of space; the commute into the city is short. However, as construction booms, it becomes increasingly important to plan for the future. The enormous challenge facing local planners is how to achieve a balance between the developments and the natural environment.

into inner valley of the Rio Grande; this approach affords a fine view of the bosque. Montaño Bridge straight ahead was built in 1996–1997 amid a great storm of opposition and controversy. (0.3)

1.8 Intersection with Coors Road (NM–448). Turn left onto Coors Road and drive north toward Corrales. Road is on an alluvial apron, just above the Rio Grande floodplain. (1.8)

3.6 Flood control dam on left. This earthen dam is part of a system of dams and channels designed to protect the area from floods coming off the mesa. (0.6)

Dragline crossing the Rio Grande to the Corrales area to begin work on the riverside drain and levee, November 30, 1930.

4.2 Passing under Paseo del Norte. The Paseo del Norte bridge over the Rio Grande was completed in 1988. (0.7)

4.9 Striking view of Llano de Sandia and the Sandia Mountain escarpment. The Llano de Sandia is a graded piedmont slope that buried former Rio Grande terraces. The highest peak with the antennae is Sandia Crest (10,678 feet). (0.2)

5.1 Turn right at light to remain on Coors Road (NM-448 to Corrales). Crossing channel of Arroyo de las Calabacillas. This arroyo system drains much of the northern Llano de Albuquerque surface, an area of over 80 square miles. Flash floods have been a problem in this area. One thunderstorm that hit the area on September 20, 1941, dropped over 20 inches of rain, resulting in an estimated 10,000 cubic feet per second of flow in the arroyo. A boulder estimated to weigh about 1,400 pounds was carried by flood waters 700 feet past the mouth of the arroyo and deposited 5 feet above the normal stream bed. No severe floods have occurred since development of the area, but they are sure to in the future. Flood-water retention devices installed here have been plagued with leakage problems due to permeability of the sandy alluvial deposits. (1.0)

6.1 Intersection with NM–528, Alameda Boulevard; continue straight ahead on 448 to Corrales. Road name changes to Corrales Road across intersection. This road descends onto the Rio Grande floodplain. Before construction of the dams upstream and before the series of drains, canals, and diversions that parallel the river were emplaced, high runoff each spring would inundate the floodplain for weeks at a time. During these times, nutrient-rich sediment, along with cottonwood and willow seeds, was deposited on the floodplain, and the flooding river would carve new channels. Although these channels are now long abandoned and filled, evidence of them still exists if one knows where to look. In many areas of the floodplain, cot-

CORRALES

This farming village was already settled when it became part of the Alameda Land Grant in 1710. It is said to have been named for the extensive corrals built by Juan Gonzalez, the founder of Alameda. There were three large haciendas then, one of which (Casa San Ysidro) still exists. Fruit, vegetables, and grapes were grown along the river, and sheep were grazed on the common grazing land that extended up the escarpment to the west and continued to the ridge above the Rio Puerco. After the American occupation of New Mexico, French and Italian families were some of the first newcomers. Vegetable farms and vineyards increased as the Albuquerque market grew. Winemaking was a mainstay in the area until Prohibition and is now enjoying a rebirth.

Corrales has fought hard to maintain its rural charm, resisting shopping malls and paving of the smaller lanes in the village. A new public library, built with donated labor and materials and with no outside funding, reflects the strong sense of community and independence.

San Ysidro Church.

tonwood and willow groves became established alongside these channels. Keep alert for aligned rows of cottonwoods that mark the edges of such old channels. (0.4)

6.5 Corrales village limits. Elevation 5,097 feet. This small community, founded in 1885, was known as Sandoval until 1966. (1.6)

8.1 Corrales Post Office. (1.0)

9.1 Down the road on the left are the historic San Ysidro church and cemetery. The existing building replaced one washed away during a flood in 1868. Some original timbers were saved and used in the new church, which was built on higher ground. In 1929 bell towers, stucco plaster, and a tin roof replaced mud plaster and a flat, earthen roof. Across from the church is Casa San Ysidro, a restored Spanish Colonial hacienda. It is managed by the City of Albuquerque; tours are available by appointment. (0.7)

9.8 MP 10. Road bends left and passes over an acequia. (1.5)

11.3 Crossing large flood control channel that diverts runoff around Corrales northward into the Rio Grande. (0.7)

12.0 Cross irrigation channel and enter Rio Rancho city limits (elevation 5,290 feet). The road begins to climb out of the Rio Grande floodplain. Prior to 1960 Rio Rancho was largely open range of the Thompson Ranch, used mainly for cattle grazing. In the early 1960s land developers removed portions of the desert vegetation and bladed a network of roads. Lots were advertised in east coast newspapers to lure potential retirees to the snowless "sunbelt." The community has become a popular residential area not only for retirees but also for others trying to escape the bustle of downtown Albuquerque.

Archaeological investigations conducted here before construction of the Rio Rancho subdivision of River's Edge revealed more than 50 sites. Thirty-five pit houses were excavated and were radiocarbon dated to A.D. 650–900. The remainder were short-term campsites where tool

The remains of the pueblo at Kuaua sit on the banks of the Rio Grande.

making and gathering of wild foods had occurred. Some of these sites were marked primarily by scatters of chipped stone and pottery, but fire hearths and food storage pits were also found. West of NM–528, recent excavations have identified 10 or more rare Archaic structures dating between 3000 B.C. and 1000 B.C. (0.7)

12.7 Intersection with Rio Rancho Drive (NM–528). Turn right at light onto Rio Rancho Drive. (1.0)

13.7 Crossing Arroyo de la Barranca. Red-colored beds of the Santa Fe Group are exposed in the distance on the left. The late Oligocene to middle Pleistocene Santa Fe Group, poorly exposed here, is the major basin-fill unit of the Albuquerque Basin. (1.3)

15.0 Views of Santa Ana Mesa and San Felipe basalt field straight ahead; Jemez Mountains in far distance at 10:00–11:00; Sangre de Cristo Mountains at 1:00–2:00; and the Sandia Mountains at 3:00. Route is on fluvial-terrace and arroyo-fan deposits related to late Quaternary intervals of Rio Grande entrenchment. (0.5)

15.5 A mile from this location is Santiago Pueblo (probably the "Alcanfor" of Spanish documents). It had long been thought that Coronado's camp during the winter of 1540–1541 was at Kuaua, until a surprising discovery was made in 1986: Roadwork along NM–528 exposed charcoal stains that, upon investigation, proved to represent outside living and work areas and shallow dugouts that contained pueblo ceramics, native plant and animal food, domestic sheep (brought by the Spanish), and—most telling—metal fragments, nails, and a riveted plate apparently from a flexible armored jack vest. Excavations in the 1930s at Santiago Pueblo recovered six crossbow bolt heads. Because Coronado's was the only expedition to carry crossbows, and because no good evidence of long-term Spanish presence has ever appeared at Kuaua, many scholars believe Coronado camped at Santiago rather than Kuaua. Six miles west of here in 1972 the Santa Fe Pacific No. 1 oil-test well penetrated 2,969 feet of Santa Fe Group deposits and bottomed in Precambrian rocks at a depth of 11,040 feet. (1.8)

17.3 Intersection with NM–550. Turn right. Bernalillo city limits. (0.6)

17.9 Entrance to Coronado State Monument on left. Turn left for Stop 4.

STOP 4 – Coronado State Monument. Situated along the banks of the Rio Grande, this site was once a flourishing village of Pueblo Indians. First settled around A.D. 1300, it is named for Francisco Vasquez de Coronado, who is thought by some to have camped here in 1540. Excavations in the 1930s revealed what many consider to be the finest examples of prehistoric painted murals in North America. The monument contains a visitor center and museum, hiking trails to the ruins of Kuaua, and restrooms. The kiva murals are displayed in a room adjacent to the visitor center. The museum was designed by John Gaw Meem, one of New Mexico's first registered architects and designer of many major buildings in the

CORONADO STATE MONUMENT

Once thought to be the pueblo where Francisco Vasquéz de Coronado and his troops spent the winter of 1540–1541, Kuaua became part of the new Coronado State Monument in 1935 in anticipation of the Coronado Cuarto Centennial in 1940. We now know that Coronado wintered near Santiago Pueblo, approximately 2 miles to the southwest. Nonetheless, Kuaua holds great importance because of the murals found in two of its kivas.

Excavations between 1934 and 1940 found three plazas, nearly 1,200 rooms, and six subterranean ceremonial chambers (kivas). Construction of the village apparently began in the late 1200s, but two major building booms occurred from 1325 to 1425 and from 1475 to 1625 or 1650. When archaeologists discovered the painted walls of Kiva 3 in February 1935, they planned to study the murals in place but soon realized that the kiva walls, built of dried balls of mud and adobe mortar, were pulling away from the earth into which they were excavated and might collapse inward at any moment. An ingenious method of jacketing the walls with tissue paper, plaster-soaked burlap strips, and lath, then encasing each wall face in wood was devised. The entire wall was then removed for transportation to the University of New Mexico. Once all the walls reached the university, an equally ingenious method was invented to remove each painted layer. The method was tedious, laborious, slow, and expensive, but the murals were removed intact. When the 2-year project was complete, they found that Kiva 3 had contained 85 layers of adobe plaster, 17 of which—usually separated by one or more blank layers—contained murals.

Many of the original murals are on display at the monument, and the kiva has been reconstructed, complete with copies of one layer of the painted walls. Although murals have been discovered in a few other kivas in the Southwest (notably Pottery Mound in New Mexico and

Inside Kiva 3 at Kuaua, 1935.

Awatovi on the Hopi mesas in Arizona), Coronado State Monument is the only place in the world where kiva murals can be viewed, thanks to the foresight and care taken at Kuaua in the 1930s.

state, and was dedicated in May 1940. Note: the monument is closed to the public on Tuesdays. **Return to NM–550 and reset trip meter to 0.0 before turning left into the highway.**

0.0 Turn left onto NM–550. (0.2)

0.2 Crossing the Rio Grande. This bridge, built in 1938, is typical of the steel girder highway bridge construction of the 1930s and is virtually unchanged. Two earthen dams were constructed upstream for flood control and water storage: Cochiti Dam, 25 miles to the north, and Jemez Dam, along the Jemez River (a major tributary to the Rio Grande). Both dams have had a significant impact on the Rio Grande by reducing the amount of sediment that the river is transporting, and by moderating the spring flows that formerly replenished the fertile bottomlands. (0.8)

1.0 Intersection with NM–313, Camino del Pueblo. Turn right to Bernalillo. From here to the end of the road log, we'll be traveling on the Rio Grande floodplain. Entering Abenicio Salazar Historical District. This section of road is part of El Camino Real. (0.9)

1.9 Sandoval County courthouse on right. (0.7)

2.6 Leaving Bernalillo. Low bluffs near interstate to the left expose reddish sandstone and conglomerate of the Santa Fe Group, capped by terrace gravels. (0.3)

2.9 Entering Sandia Pueblo. From here you have a view of some irrigated farms that share the floodplain with the cottonwood–willow bosque ecosystem. (2.3)

5.2 MP 4. The low ridge that juts out from the Sandia Mountains is Rincon Ridge, composed entirely of Precambrian metamorphic and granitic rocks riddled with dikes. At 9:00, the highest peaks of the Sandia Mountains are home to radio and television towers. Note elevated railroad tracks on left, designed to be protected from flooding. The last flood to submerge this area was in 1941. The largest historical flood of the Rio Grande happened in 1828 and was so large that it inundated the entire valley from Albuquerque to El Paso, Texas. With the completion of Cochiti Dam in 1975, the threat of flooding in the Albuquerque area has been greatly reduced. (0.3)

5.5 Sandia Pueblo is located among trees on the left. (1.8)

7.3 The Sandia Pueblo archaeological sites are to the east. (0.7)

Silkscreened reproduction of the murals at Kuaua by Martha Walker, School of American Research, 1964. A number of these murals are now on display in the museum at Coronado State Monument.

BERNALILLO

One of the earliest districts in New Mexico to be settled by Spanish landholders, La Angostura de Bernalillo (a few miles north of town) held several estancias by the 1650s and 1660s. In 1695, after the Pueblo Revolt, Governor Diego de Vargas established the town of Bernalillo. For much of the 17th, 18th, and 19th centuries, Bernalillo was a major mercantile center. In 1880 the Atchison, Topeka, and Santa Fe Railroad made Albuquerque the location for its shops and switching yards, and the town began to subside into the quiet village it is today.

Bernalillo is one of a small group of Hispanic towns and Indian pueblos that still dance Los Matachines, apparently brought to New Mexico by Spanish settlers soon after 1598. A blend of dance and drama, the presentation is held on August 10 during Bernalillo's Fiesta of San Lorenzo. Neither the origin of the name Matachin nor the exact meaning of the performance is clear, but there is agreement among researchers

Our Lady of Sorrows church in the Abenicio Salazar Historical District, Bernalillo, ca. 1940.

that the dance contains both Old World and New World linguistic, costuming, and dance formation components. It may be that the dance stems partially from an Aztec dance known as netotiliztli. In New Mexico the performers include a Monarca and 10–16 or more Matachines (including four capitanes), all of whom wear masks and ribboned headdresses and carry three-pronged tridents and rattles. La Malinche, the only female, is a young girl dressed in white shoes, white stockings, white dress, and white veil. El Abuelo, the grandfather clown, who directs the Malinche and acts as master of ceremonies, wears old clothes, a mask, and carries a whip. Usually a young boy acts as El Toro, transformed into a bull by an animal skin and a horned headdress. The musical accompaniment is always played on violins and guitars, but drums and vocal chants are added at some pueblos.

The Abenicio Salazar National Historic District, formally listed on the National Register of Historic Places in 1980, was named to honor the work of this early 20th century craftsman in adobe architecture. One of his many buildings, Our Lady of Sorrows High School, has been restored by the community and his descendants. Nearby, on the west side of the street, is the graceful Our Lady of Sorrows Church (above), built in 1857. The church was abandoned in 1971, but parishioners and community members continue to work, against great odds and with little money, at stabilization and restoration.

8.0 Sandia Lakes on the right are artificial ponds stocked for fishing. (1.1)

9.1 MP 0. Crossing 2nd Street. Keep right and continue straight. (0.3)

9.4 North Diversion Channel is designed to divert floodwaters from arroyos draining the Sandia Mountains upstream to empty into the Rio Grande north of Albuquerque. (1.6)

11.0 Turn right at Alameda Boulevard (NM–528). (1.0)

SANDIA PUEBLO ARCHAEOLOGICAL SITES

Twenty-six archaeological sites have been identified in this 2-mile stretch. All of the sites are clustered along the gravel tongues in a swath only 0.3 mile wide. A close look at the gravels suggests the reason: these are not just limestones or crumbs of decomposed granite washed down from the Sandias. Instead, they are a combination of well-traveled quartzite cobbles from as far north as Colorado (some of the same rocks appear in the San Juan River), obsidian and basalt cobbles from the Jemez Mountains, chert cobbles from several possible upstream sources, a few cobbles of red sandstone, and chunks of petrified wood.

Nearly all of these rocks had important uses for prehistoric people. Obsidian, chert, and fine-grained basalt were excellent raw materials for stone tools including projectile points for spears, darts, and arrows used in hunting; knives and scrapers for skinning and butchering animals, making wood and bone tools, and preparing vegetable foods; burins for engraving and scoring wood and bone; punches for piercing bone and leather; drills for wood, bone, stone, shell, and turquoise; and spokeshaves for straightening and smoothing arrow and dart shafts. Mauls, axes, and choppers were made from tough quartzite and basalt cobbles. Small flattish quartzite cobbles made perfect one-hand "pillbox" manos for grinding seeds, but small vesicular basalt cobbles were even better, since they never required sharpening. Desert Archaic people used quartzite cobbles in their roasting pits to provide heat for long, slow cooking (dozens of these cooking pits have been found on the Sandia gravel tongues). Quartzite cobbles even had an architectural use. The Basketmaker III–Pueblo I people who built their houses on these quick-draining tongues were probably never forced to slosh around ankle-deep in water, as were those who dug their houses in alluvial soils.

Archaeological sites on the Sandia gravel tongues help to provide a picture of prehistoric life along the Rio Grande. After Basketmaker III–Pueblo I, no one ever lived on the gravel tongues again (later residential sites were built down on the floodplain), but people visited frequently to hunt, gather, and knap stone.

Obsidian point from Frijoles Canyon.

12.0 Turn left onto south Rio Grande Boulevard (NM–194). (0.4)

12.4 We are paralleling the Rio Grande on its floodplain. Note flood control levee at 3:00. On the river side of the levee are large, scattered steel structures called Kellner jetty jacks. Since 1951 the U.S. Bureau of Reclamation has installed over 100,000 jacks to protect the levee. The jacks were designed to retard flood flows by trapping sediment and promoting the establishment of vegetation. In places, the jacks are now being removed. (0.1)

12.5 Entering Los Ranchos de Albuquerque, now an exclusive residential area but once an old Spanish Colonial village. Home builders on the floodplain must be wary of soils with a high clay content. These soils can expand and contract upon wetting and drying, causing expensive foundation damage. (0.5)

13.0 Crossing over Paseo del Norte. (2.0)

15.0 Chavez Avenue. North of Chavez Avenue and east of Rio Grande Boulevard may be the location of the old Los Ranchos plaza, dating from 1750 or earlier. Test excavations in 1996 revealed foundations of adobe walls, a possible blacksmith's shop, historic Pueblo and Euro–American ceramics, and enormous amounts of domesticated animal bone (cow, pig, goat, sheep, burro, and horse). A previous property owner reported many mounds (one of them 100 feet long) of collapsed adobe structures and a large rectangular area paved with cobblestones. Both test excavations and historic documents tell of repeated flooding in the North Valley area.

Historical records show that Los Ranchos and four other plazas (Los Duranes, Los Candelarias, Los Griegos, and Los Poblanos) were settled along the river branch of El Camino Real north of Albuquerque about 1750. Another, Los Gallegos, was settled about 1785. Los Ranchos, established on the Elena Gallegos Grant, was the largest of these villages. After Mexican independence from Spain in 1821, it functioned as a judicial center, and, during New Mexico's days as a U.S. Territory, served as the county seat for Bernalillo County from 1851 to 1854. Though still present as North Valley street names, the other old plazas have disappeared inside Albuquerque. In 1958, anticipating the same fate, Los Ranchos residents incorporated as Los Ranchos de Albuquerque. **Watch for sharp bend ahead.** (0.9)

15.9 Alexander Valley Vineyards. New Mexico had a thriving wine industry prior to Prohibition. In spite of Prohibition, it continues to thrive today. (0.3)

16.2 Montaño Boulevard You are now in an old growth stand of the most common

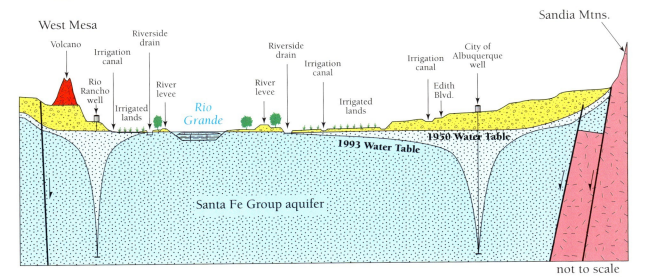

Cross section of the Rio Grande valley showing the modified water table, irrigation canals, riverside drains, and flood control structures.

ALAMEDA PUEBLO

Alameda Pueblo, a large village on the west bank of the Rio Grande, was built around A.D. 1200. It was still occupied when the Spanish arrived in 1540. The river channel shifted in 1735–1736, placing the pueblo on the east bank. Spanish documents indicate that the pueblo was abandoned during the Pueblo Rebellion of 1680, and only a few people agreed to move back when the Spanish government attempted to resettle the pueblo in 1702. By 1710 resettlement was clearly a failure, and the 50 inhabitants went to live with other southern Tiwa speakers at Isleta Pueblo south of Albuquerque. In that same year, Governor José Chacón issued the Alameda Pueblo Grant to Captain Francisco Montes Vigil in recognition of his military service in the 1692 reconquest of New Mexico. He subsequently sold the grant to Juan González, who founded the village of Alameda about 1712.

The ruins of the pueblo itself lay undisturbed until 1916, the mound standing 10 feet high and covering 3–4 acres, when much of the mound was destroyed to provide fill for county roads. Nearby residents told a researcher in 1931 that "a great many human skeletons...and a great quantity of pottery [were] brought out" and remembered "many walls and rooms." The Alameda Elementary School was later constructed over part of the nearly obliterated site.

The Highway 66 bridge over new Alameda interior drain along the Rio Grande in 1930.

The riverside drain, which sometimes contains more water than the river itself, contains an ecosystem of its own. Dragonflies, water striders, mosquitoes, and other insects live among the aquatic flora such as watercress, cattails, and willows. Toads, frogs, snakes, muskrats, and birds complete the food chain.

cottonwood along the Rio Grande, the Fremont cottonwood (Populus fremontii var. wislizenii). These trees are about 80 years old and have nearly reached their mature height of about 50–60 feet. Under ideal conditions, cottonwoods can grow to over 90 feet. Cottonwood trees produce male and female flowers on separate trees. Mature female seeds are covered with a white fuzzy cotton that gives the trees their name and permits wide dispersal of seeds by the wind. Cottonwoods were a welcome sight for early travelers through the Southwest, as they grow where there is water, and they provide welcome shade from the hot summer sun. Cottonwoods also provide abundant habitat for birds and other wildlife. Other cottonwoods found along the Rio Grande include the Plains cottonwood and the narrowleaf cottonwood. The leaves of the Rio Grande variety are broad and more coarsely toothed than the Plains cottonwood; leaves of the narrowleaf cottonwood are more willow-like. All cottonwoods are members of the willow family. (0.4)

16.6 Crossing Griegos Road. The Los Griegos Historic District on Griegos Road between Rio Grande and Guadalupe Trail NW marks Los Griegos plaza, settled by the Griegos family in the 1750s. Prepare to turn right ahead. (0.8)

17.4 Turn right onto Candelaria Road and follow signs to Rio Grande Nature Center. (0.8)

18.2 STOP 5. Rio Grande Nature Center. Park car and walk to visitor center. The 270-acre Rio Grande Nature Center preserves the most magnificent remaining stand of Rio Grande cottonwood bosque in the world. For decades the river was viewed only as a water source for farms, and a flooding threat to be tamed. As management agencies continued to channelize, drain, and reclaim the river, thousands of acres of wetlands were lost. In 1969 the U.S. Bureau of Reclamation proposed clearcutting the bosque to enhance irrigation. Opposition to this short-sighted plan was based on an appreciation of the river's educational, ecological, recreational, and aesthetic values. By 1976 the city had purchased 177 acres, and in 1980 the state leased 38.8 acres for the visitor center, a pond and wetland area, and a foot trail through the bosque to the river. In 1983 the state legislature authorized the Rio Grande Valley State Park, which provides protection for the bosque from the northern to southern boundaries of Bernalillo County.

The pond attracts a variety of migratory waterfowl, who feast on such delicacies as sago pond weed, deep water duck potato, elodea, ducks meat, three square rush, hard stem bullrush, and cattails. A visit to the pond may reward you with a glimpse of turtles, muskrats, snakes, redwing blackbirds, grebes, kingfishers, ducks, geese, and

The pond at Rio Grande Nature Center.

perhaps a heron. Gambusia fish prowl the water in search of mosquito larvae.

The presence of the bosque ecosystem is intimately related to the water table. If the water table drops below the level at which a lake can be maintained, a marsh develops. As the level continues to drop, a salt grass meadow appears, salt grass replaces the marsh, and small willow species begin to grow. Finally, at a critical level, the tall willows, cottonwoods, Russian olive, and salt cedar dominate. If the water table were to drop quickly below a certain depth, the cottonwoods would become stressed, and parasites such as mistletoe may become established. Eventually the trees would die, and the bosque would become desert scrubland. **Retrace route back to Rio Grande Boulevard and turn right to return to I–40.**

SCENIC TRIP FIVE - (24 MILES)
The Albuquerque Volcanoes

This short trip to the West Mesa offers a hands-on look at the Albuquerque volcanoes, geologically young features that have now been preserved as part of Petroglyph National Monument. Driving time for this trip is about 2 hours, but we recommend a half day to explore the volcanoes. There are no facilities at the site and little shade, so be sure to bring water, food, sturdy shoes, hat, and sunscreen. The Volcanoes Day Use Area is gated and open only from 9 a.m. to 5 p.m. each day; it is closed during inclement weather. For more information call Petroglyph National Monument at (505) 899-0205. Access to this area is via Paseo del Volcan, which can be reached by taking Exit 149 from Interstate 40.

The position of the Albuquerque volcanoes is closely aligned with and related to the County Dump fault. Examination of soil surfaces that are offset show that this north-striking, nearly vertical fault was intermittently active in middle to late Pleistocene time. However, no displacement has occurred in the last 20,000 years. The volcanoes themselves erupted from fissures aligned along a tension fracture that parallels the County Dump fault.

On January 4, 1971, an earthquake with intensity VI was centered in this area; an aftershock of intensity III followed. The ground tremor was felt over an area of over 600 square miles. Minor damage, consisting mostly of cracked plaster, broken windows, and damage to fallen objects was reported in the west and northwest parts of the city. The University of Albuquerque suffered

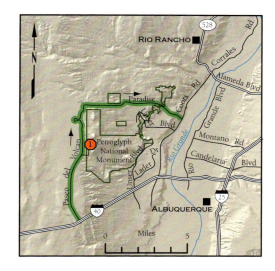

damage when corrosive chemicals in the chemistry lab fell to the floor and when thousands of books flew off library stacks. The gymnasium at West Mesa High suffered structural damage, a Persian gazelle died at the zoo, and a prisoner was shaken out of his upper bunk at the county jail. The quake appears to have been related geographically to the County Dump fault.

Exit I-40 at Exit 149, Paseo del Volcan, and follow the off-ramp to intersection and stop sign. Reset trip meter at this stop sign.

0.0 Stop sign at foot of off-ramp. Turn right onto Paseo del Volcan, heading north toward Double Eagle II Airport. You are driving on the Llano de Albuquerque. This surface represents the highest level that the basin filled at the

Rattlesnakes are often seen in this area; use caution when hiking.

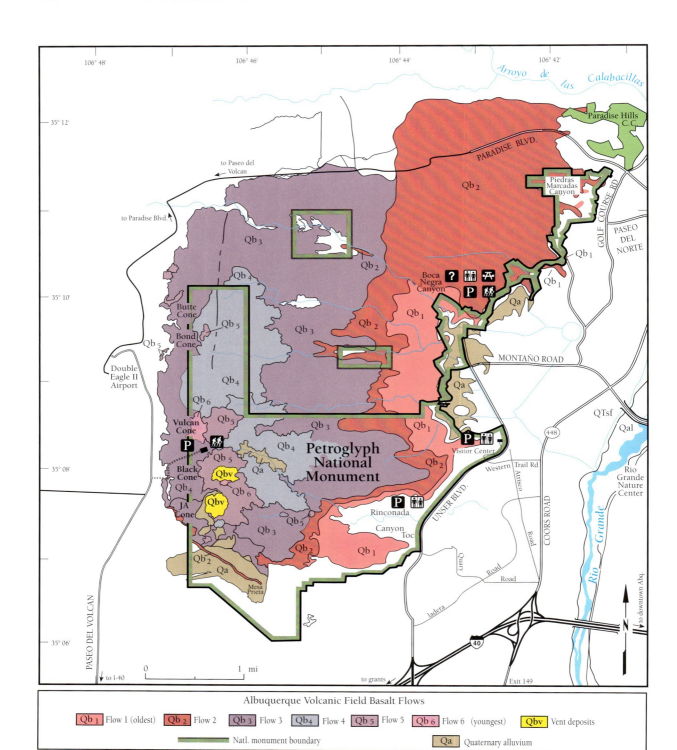

end of Santa Fe Group deposition. (0.4)

0.4 Double cattle guard. Vegetation includes various grasses, snakeweed, yucca, and prickly pear cactus. (1.2)

1.6 Double cattle guard and powerlines. (1.0)

2.6 Mountains surround us: Mt. Taylor (elevation 11,389 feet) at 9:00; Nacimiento Mountains straight ahead; Jemez Mountains at 1:00; Albuquerque volcanoes at 1:30; Sangre de Cristo Mountains at 2:00; and Sandia Mountains at 3:00. (1.8)

4.4 Shooting Range turnoff on left; continue straight. Prepare to turn right ahead onto unmarked gravel road. Two miles west of here in 1983 Shell Oil Company's West Mesa Federal No. 1 well penetrated over 8,500 feet of Santa Fe Group deposits and bottomed in Jurassic bedrock at a total depth of 19,377 feet. (0.4)

4.8 Turn right onto gravel road. (0.3)

5.1 STOP 1. Volcanoes Day Use Area, Petroglyph National Monument. Be sure to lock your vehicle and beware of rattlesnakes. Rock collecting is not allowed in the park.

A far-sighted woman named Ruth Eisenberg is singularly responsible for the preservation of these volcanoes. In the late 1960s, while undertaking a University of New Mexico (UNM) class project researching the ownership of the volcanoes, she discovered that they were privately owned, and for sale. Fearing that they would be subdivided and suburbanized, she established "Save the Volcanoes." Through hard work and perseverance she managed to get public and political backing for her visionary project. In 1973 the city bought the three southern cones and (in 1976) the two northern ones. The old Volcano Park was later absorbed into Petroglyph National Monument.

There are five large cones and at least 10 smaller ones aligned along a north–south zone of fissures paralleling the County Dump fault in this area.

OPPOSITE: **Geologic map of the Albuquerque volcanoes, showing the relative ages of the individual flows.**

Aerial photo of the Albuquerque volcanoes.

Approximately 155,000 years ago, these volcanoes erupted in fountains of glowing lava and ash that may have reached heights of 30 feet or more. The eruptions were small and not especially violent: had people been living in the area then (they were not), they could probably have watched the eruptions from a distance of a half mile or less. The first phase of eruptions occurred along the entire length of the fissure, but later eruptions became concentrated at local hot spots. The cones probably represent the final eruptive phase. Individual eruptions may have lasted for a few months at a time.

sharkskin, corded, festooned, filamented, shelly, and slab.

Mineralogically these lavas are similar to the lavas of Hawaii and some basalt fields in the Rio Grande rift such as the Taos Plateau volcanic field. Hike to the top of Vulcan cone along a rocky trail for a spectacular view of the basin. Vulcan

The three prominent cones of the Albuquerque volcanoes rise just above the basalts of West Mesa.

Lava mainly flowed eastward toward the Rio Grande at speeds of perhaps 5–10 miles per hour. About six separate flows covering an area of 23 square miles are recognized in the field. There are two types of lava flows here, pahoehoe and aa, both Hawaiian terms. Pahoehoe flows have a smooth or ropy surface indicating fluid flow (lower viscosity), whereas aa has the rough, jagged, blocky surface of a more viscous lava. Pahoehoe flows can further be divided by wonderfully descriptive terms such as elephant-hide, entrail,

(named for the Roman god of fire) is the tallest cone and stands about 200 feet above the mesa. Notice the different textures and colors of the lava rock. The white coating is the mineral calcium carbonate, produced by weathering processes long after the lava solidified. The dark-gray to reddish-brown basalt contains white plagioclase and rare dark-green olivine crystals. Holes are vesicles produced by gas escaping from the lava as it cooled. Granular materials, typically in layers, are cinders that were forcibly ejected from the volcano. In many places, you can see flow structures and globs of lava that spattered on the rocks. Most of the flows here are pahoehoe.

From the summit of Vulcan the linear chain of cones is obvious. To the south are Black and JA cones, and to the north are the smaller Bond and Butte cones. Earlier in the century there was a sixth cone named Cinder, but it was quarried for its black and red landscaping cinder; all that remains today is an area of dark rubble. Along the south flank of Vulcan is a second quarry where cinders were mined for use in cinder blocks. The lava flows continue to the east for about 2 miles. Distant views include the Nacimiento and Jemez Mountains to the north, Mt. Taylor to the west, and Ladron Peak and the Magdalena Mountains to the south. The Jemez Mountains and Mt. Taylor are also volcanoes, but they are much larger and older and erupted more violently than the Albuquerque volcanoes.

Hike down the flank of Vulcan to view the quarry. At least two parasitic cones formed low on the flanks of Vulcan. Be careful while walking near the quarry edge as loose material may break away. The order of layering yields the eruptive history of the cone; alternating layers of cinders and thin flows show that the cone formed from a series of periodic cinder explosions and eruptions of lava.

Hike south to Black and JA cones, a distance of about 1 mile, to see how they compare to Vulcan. Black cone has been almost entirely removed by mining, as it consisted mainly of cinder. Like Vulcan, the JA cone consists mostly of layers of spatter and cinder. **Return to Paseo del Volcan. Set trip meter to 0.0**

Eruption of the Albuquerque volcanoes 100,000 years ago likely resembled this modern fissure eruption in Hawaii.

0.0 Turn right on Paseo del Volcan to continue northward on loop that ends at Coors Boulevard via Paradise Hills. (0.6)

0.6 Vulcan cone at 3:00. (0.6)

1.2 The road is on the western edge of a basalt flow that probably originated in the Vulcan cone area. (0.5)

1.9 Double Eagle II Airport on left. Completed in 1983, this airport is named in honor of the hot air balloon that made the first successful crossing of the Atlantic Ocean. The pilots, Maxie Anderson, Ben Abruzzo, and Larry Newman, were Albuquerque residents. The Double Eagle II balloon and gondola now reside at the Smithsonian Air and Space Museum in Washington, D.C. (0.4)

2.3 Stop sign. Turn right on Paseo del

WILD MAGNETIC FIELDS

These volcanoes produced a remarkable geologic surprise in 1990 when Dr. John Geissman, a paleomagnetist from the University of New Mexico, sampled the basalts for a geomagnetic study. Geologists are able to study magnetism in rocks such as these because, as lavas cool, certain minerals line up like tiny compasses in the earth's magnetic field. He discovered that the basalts preserved an unusual shift in the earth's magnetic pole.

Although compasses today point to the north magnetic pole, at the time that the volcanoes erupted, compasses would have pointed eastward! Seven such shifts have been found in the geologic record over the past 730,000 years. Each of these wild magnetic events may have lasted a thousand years or so. Normally, the earth's magnetic field behaves as if a giant, long bar magnet existed within the planet. During these events, however, the field acts as if the magnet was breaking into several smaller magnets. The most widely accepted theory explaining the Earth's magnetic field states that currents of molten rock in the core produce an electric field, which, in turn, produces the magnetic field.

Scientists do not understand the driving mechanism behind these excursions. One possible explanation for the magnetic shift recorded in these basalts is that a large meteorite hit Earth with enough force to alter the electric current flow in the core, thus causing the magnetic excursion.

One question has been the effect of magnetic reversals on Earth's life forms and on the process of evolution. Earth is shielded from much of the dangerous solar radiation by the Van Allen belt and similar belts of charged particles in orbit around the planet. A zero magnetic field would probably eliminate the belts, allowing high levels of charged particles to reach the biosphere. Such radiation may kill certain sensitive species, such as plankton and amphibians, and increase mutation rates, cancer, and visual damage in other species. It's also thought that some animal species, such as birds and insects, navigate by magnetic fields. A reversal today would certainly affect compasses and eliminate long-distance radio communications for a few hundred years. Perhaps the biggest change could relate to climate, but the interactions between charged particles and the greenhouse effect are poorly understood.

Norte. Water tank on left. Depth to ground water here is more than 500 feet. The lack of adequate water supplies in this area has been a major hindrance to development. This area is underlain by Holocene alluvial and eolian deposits. Vegetation consists of a variety of grasses, snakeweed, yucca, and prickly pear cactus. (0.5)

2.8 Paralleling the road on the right is the northernmost string of volcanic cones. The two prominent ones are Bond to the south and Butte to the north. (0.9)

3.7 Road curves right; high-power electric lines just ahead. The road begins a gradual decent into an unnamed drainage and leaves the Llano de Albuquerque surface. This drainage eventually joins the Rio Grande about 5 miles to the east. It is only during the most intense rainstorms that water reaches the river. Most of the time, the water seeps underground or evaporates before reaching the river. (0.8)

4.5 Good view of Nacimiento Mountains at 11:00 and Jemez Mountains on the far horizon at 1:00. Eight miles to the northwest in 1976 the Shell Santa Fe Pacific No. 3 oil well penetrated 3,996 feet of Santa Fe Group deposits and bottomed in Triassic strata at a total depth of 10,276 feet. (1.0)

5.5 Spectacular view of Sandia Mountains ahead. A thin layer of alluvial and eolian sediments covers the Qb_3 lava flow (see geologic map) in this area. Small hill on right consists of highly vesiculated basalt that may represent a small isolated vent. (0.4)

6.0 About 0.5 mile to the left is the northernmost extent of the flows. Note scattered juniper trees growing in arroyos on both sides of the road. Just ahead, the road traverses an alluvial apron of sediments that overlies Santa Fe Group deposits. (1.4)

7.4 Cross Rainbow Boulevard and the entrance to Ventana Ranch, a residential development. This area was developed in the middle to late 1990s. Water is supplied to these homes from deep wells. (0.4)

7.8 Turn left onto Universe and make the next right onto Paradise, which will lead you back to Coors Blvd.

9.0 Paradise Hills Park on left. We are traveling on the Segundo Alto terrace surface. (0.4)

9.4 Cross Lyons Road. Begin descent into inner valley of Rio Grande. Be prepared to make a series of turns to eventually meet Coors Boulevard. To return directly to Albuquerque, stay on Paseo del Norte and cross bridge. (1.2)

10.6 Cross Golf Course Road. Prepare for right turn ahead. (0.8)

11.4 Turn right on Eagle Ranch Road and then left on Paseo del Norte. Stay in right lane to reach Coors Boulevard. (0.3)

11.7 Keep right and exit to Coors Boulevard. Just ahead you can turn right to Albuquerque via Coors Boulevard or turn left to continue Trip 4 at Mile 4.2.

SCENIC TRIP SIX - (82.6 MILES)
Rio Grande Rift to the Colorado Plateau

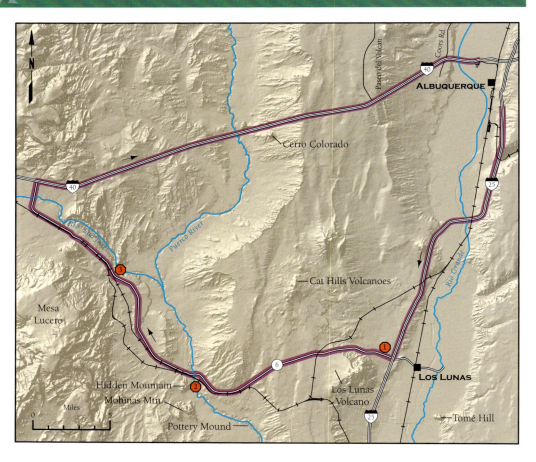

This trip explores the sparsely populated region southwest of Albuquerque. It begins by heading south on I–25 along the Rio Grande toward Los Lunas to view exposures of the Santa Fe Group and a variety of volcanic features. An optional stop can be made at the Pueblo of Isleta. At Los Lunas we depart the river and travel west on NM–6 past the Los Lunas volcano and over the Llano de Albuquerque to the Rio Puerco. NM–6 was part of the route of U.S. Highway 66 from 1927 to 1937. Along this route we will pass Pottery Mound, an archaeological site excavated by the University of New Mexico in the 1950s. After crossing the Rio Puerco, the trip heads northwestward, crossing from the Albuquerque Basin to the Colorado Plateau with its multicolored rock exposures and picturesque mesas and buttes. I–40 is followed back through the Laguna Pueblo tribal lands and into Albuquerque.

There are few facilities along the trip route. You will be traveling through land that has not changed much since the first Spanish explorers passed by in 1540. Be sure to have a full tank of gas, lunch, and water. Non-stop driving time is about 3 hours, but it is best to take a full day. Much of the land along the route is private or reservation land, so please do not venture far from the roads. Log begins heading south on I–25 at Gibson Boulevard.

0.0 MP 223. Crossing Gibson Boulevard. Drive south on I–25. Route ahead is on a Holocene alluvial-fan apron that extends westward from the base of the escarpment (9:00–11:00) that forms the outer rim of the Rio Grande valley. The escarpment ascends to the left to a nearly level surface on which the Albuquerque International Airport is built. This surface is a remnant of an older Albuquerque Basin floor and is underlain by upper Santa Fe Group sediments. (0.5)

0.5 Sunport Boulevard. I–25, from the Colorado border to Las Cruces, is the longest highway in the state at 462 miles. (0.5)

1.0 MP 222. Albuquerque International Airport at 9:00. (0.3)

1.3 University of New Mexico South Golf Course at 9:00. The area just west (to the right) of the freeway is part of the South Valley Superfund site. During the 1950s and 1960s, this region became an industrial center handling metal-parts manufacturing, organic chemicals, and petroleum products. The soil and ground water were subsequently found to be contaminated.

The Environmental Protection Agency (EPA) designated this area as a Superfund site in 1982, making it eligible to be cleaned using funds generated from taxes on companies who manufacture oil and 42 specific hazardous chemicals. The EPA has removed 3,450 gallons of oil and 63,580 pounds of contaminated soil and debris, and 20 private and two municipal water wells have already been condemned. (0.7)

2.0 MP 221. Slopes to the east are composed of upper Santa Fe Group sand and gravel, with some pebbles derived from source areas north of the Albuquerque Basin (e.g., pumice and obsidian from the Jemez Mountains). Local lenses of lacustrine (lake) clay, mud, and sand are also present, and sets of cross-strata dip mainly southward. These features suggest that the upper Santa Fe sediments in this area accumulated in a basin-floor environment of channel and overbank deposits of the ancestral Rio Grande.

At road level are sediments derived from the northwest that were deposited by the ancestral Rio Puerco. (0.3)

2.3 Crossing Rio Bravo Boulevard. (0.3)

2.6 Abandoned South Broadway city/county landfill is located behind the fence at 9:00. The landfill opened in 1983 for commercial and household refuse and closed in 1992. Federal laws require several years of post-closure monitoring for methane gas release and ground-water contamination. Long-term monitoring is especially important here, because the landfill rests on permeable deposits that overlie a fairly shallow water table. (0.4)

Aerial view of the south valley in 1935. Note former river course to the west of the modern Rio Grande. An even older river meander is preserved in the lower left.

Aerial view of the south valley in 1996. Most of the farmland has been replaced by urban sprawl, the old river course is indistinguishable, and the older river meander is now farmland. The Rio Grande is confined to a floodway.

3.0 MP 220. Begin descent into valley of Tijeras Arroyo. Good view of inner valley of Rio Grande with Mt. Taylor in far distance at 3:00. (0.4)

3.4 Crossing railroad spur. Walls of the lower Tijeras Arroyo valley are underlain by ancestral-river deposits of the upper Santa Fe Group. A fascinating assemblage of vertebrate mammal fossils was recently discovered and studied on the south side of Tijeras Arroyo near the top of the scarp at 9:00. (0.5)

3.9 Bridge over Tijeras Arroyo. The drainage channel can become a raging torrent after heavy mountain rains. The headwaters of the Tijeras drainage system lie about 20 miles eastward in the Sandia and Manzanita Mountains. Because this system drains a large and well-watered area, the Tijeras Arroyo is one of the few drainages in the Albuquerque area that extend from the mountains all the way to the Rio Grande. The channel empties into the river about 2 miles west of here. (0.7)

4.6 Roadcuts ahead are in upper Santa Fe Group ancestral-river sediments. (0.7)

5.3 Magdalena Mountains (west of Socorro) on the distant skyline at 12:30, Los Lunas volcano and Sierra Ladrones at 1:00, and Isleta volcanic center at 1:30 across Rio Grande floodplain. (0.3)

5.6 Our route descends onto a Holocene alluvial-fan apron graded to the approximate level of the present Rio Grande floodplain. Between here and Broadway we will cross a large, down-to-the-west fault. This fault has altered the course of ground water in the sediments, and wells near here pump warm water. The mesa to the east, Mesa del Sol, is the site of a future planned community of 80,000 people. A test well there penetrated 1,700 feet of ancestral Rio Grande sediments. (1.4)

7.0 MP 216. Broadway interchange ahead. Continue south on I–25, which bends to the right. (1.5)

8.5 Crossing main irrigation canal. The forest along the banks of the Rio Grande ("the bosque") consists mainly of cottonwood, willow, salt cedar, and Russian olive. The mounds on both sides of the river are levees built to keep flood waters within the channel. The ditches and drains along the river provide important habitat for establishment of salt cedar and Russian olive, both exotic, invasive species. Just ahead we'll cross the active channel of the Rio Grande. If the river level is low, the braided character of the river will be evident. This is the longest highway bridge in the state (2,314 feet long). (0.6)

9.1 Crossing Isleta Boulevard and entering Isleta Indian Reservation. Ahead the highway slashes a large roadcut through Black Mesa. Note the crossbedded gravel and sand of the upper Santa Fe Group deposited by the ancestral Rio Grande. The crossbedding reveals much to the sedimentologist about

THE SIERRA LADRONES FORMATION

The Santa Fe Group in Tijeras Arroyo is represented by the Sierra Ladrones Formation, which was deposited by the Rio Grande about 1.6 million years ago. Well-preserved bones and teeth of mammals including *Glyptotherium* (giant, heavily armored armadillo-like creatures that originated in South America and became extinct during the ice ages), *Hypolagus* (early rabbit), several varieties of *Equus* (horses), *Camelops* (camel), and two types of *Mammuthus* (mammoths) have been found in Tijeras Arroyo. From these fossils and the sediments in which they were found, researchers have concluded that during the early Pleistocene this area was a semi-arid piedmont plain covered by open grassland or short-grass prairie, with low-sinuosity, braided streams similar to the modern Rio Grande.

Camelops

sedimentary environment and direction of river flow. The basalt flow visible in the roadcut thins from 40 feet to zero. The source of the mesa-capping basalt flow (2.68 million year old) is a puzzle, because it has no outcrop connection to the Isleta volcano, its chemistry is different from the volcano, and it overlies ash from the volcano. A volcanic vent may lie buried to the southeast beneath the Rio Grande floodplain. (0.7)

birds, turtles, muskrats, fish, gophers, beavers, and other fauna. The maintained roads along the ditches are also historically important recreational avenues for valley residents who choose to walk, run, bike, or ride horses. The canals and ditch bank roads are maintained by the Middle Rio Grande Conservancy District. Notice the burned cottonwoods in the bosque area to the left. Human-caused bosque fires kill cottonwoods and willows and allow the

Roadcut at mile 9.1 reveals crossbedded sediments of the Sante Fe Group overlain by basalt.

9.8 In roadcuts to left and right, sand and gravel of the ancestral Rio Puerco are overlain by basaltic tuff (compacted ash) erupted during early development of the Isleta volcano. (0.5)

10.3 Crossing Coors Road. Basalt of Black Mesa overlies tuff in roadcuts ahead. (0.2)

10.5 Interstate begins ascent onto basalt flow; Isleta volcano is straight ahead. This 300-foot-high volcano is a compound feature formed from five basalt flows. The base of the volcano is within an earlier maar crater that is almost completely buried except on the northeastern and eastern sides. A maar is a special type of low-profile volcano formed by steam and ash explosions rather than magma eruptions. The rim of the maar is seen to the right where ash beds change from dipping gently northeast away from the volcano to dipping steeply southwest into the vent. The lowermost flows of the later volcano, dated at 2.75 ± 0.03 million years, may have been part of a lava lake that erupted in the maar. The second flow above the maar has been dated at 2.78 ± 0.06 million years. (0.5)

11.0 MP 212. Contact of basal basalt flow (lava-lake unit) on tuffs of maar exposed in valley wall to right. Note intratuff unconformity with tuff and breccia of Isleta maar on truncated, flat-lying tuffs deposited outside the crater. This contact is the depositional rim of the maar. For the next 0.4 mile, thin upper Santa Fe and

ISLETA PUEBLO

Isleta Pueblo is located to the east between the river and the freeway. The present pueblo, a southern Tiwa-speaking village, is an amalgamation of several earlier farming villages on both banks of the Rio Grande. Like the other pueblos, Isleta tradition records a northern ancestry, but the people of Isleta also believe that one group of ancestors came from the south. For unknown reasons, Isleta did not take part in the massacre of Spanish settlers during the Pueblo Revolt in 1680. Refugees from Santa Fe found Isleta deserted as they fled south. A year later, when Governor Otermín attempted a futile reconquest of New Mexico, he attacked a reoccupied Isleta, taking 511 captives and burning the village. The captives were taken to El Paso and resettled, where some of their descendants live today at Ysleta del Sur, a community now surrounded by the city of El Paso.

Once largely agricultural, Isleta was particularly famous for its vineyards and wine making (until Prohibition put an end to it). Although ranching and farming continue, many Isletans today are employed at Sandia National Laboratory, Kirtland Air Force Base, and by private businesses in Albuquerque.

post-Santa Fe terrace deposits overlie the flow. (0.8)

11.8 Outlying basaltic tephra exposed in the roadcut to the right. About 1.5 miles west of here is the site of the deepest oil-test well drilled in the basin. It is the 21,266-foot-deep Isleta No. 2 oil test completed in 1980 by the Shell Oil Company. The well penetrated only Cenozoic units, including 14,460 feet of the Santa Fe Group. The well was dry and was abandoned. The main community of Isleta Pueblo, including the historic church and plaza, is about 1.5 miles to the east on a basalt-capped bench above the Rio Grande floodplain. (0.3)

12.1 Cone of Isleta volcano at 3:00; Isleta water tanks at 9:00. Spectacular exposure showing maar crater rim marked by anticlinal fold. Roadcut exposes thinly bedded, light-colored ash layers overlain by cinders and capping basalt flow. The brick-red contact between cinders and basalt flow is a "bake zone" formed when the hot basalt flowed over the cinders.

About 5 miles east of here, on Mesa del Sol, the Transocean Isleta No. 1 oil-test well in 1978 was drilled to a total depth of 10,377 feet, bottoming in Precambrian rocks. Interestingly, the Santa Fe Group was only 5,039 feet thick. When compared to the 14,460-foot thickness in the Isleta No. 2 well, it shows that the Santa Fe Group thickens by almost 10,000 feet between the two wells in a distance of only 8 miles. Thus, a fault (or series of faults) with over 9,000 feet of movement lies between the two wells. Because ash from Isleta volcano is found in the fault zone at the same elevation as the volcano, much of the fault displacement must be much older. (0.8)

13.0 MP 210. Isleta Pueblo interchange. Continue south on I–25. Route for next 7 miles is on an inset river terrace that is about 140 feet above floodplain level. This surface probably is correlative with the late Pleistocene Segundo Alto that we crossed on Trip 3. (0.8)

13.8 Panoramic view across southern Albuquerque Basin includes Los Lunas volcano at 1:00, Mesa Lucero on distant

Construction of the Isleta diversion dam across the Rio Grande on October 3, 1933. This dam is one of several in the middle Rio Grande basin that move water into the irrigation canals.

Fighting a flood at the Chical break near Isleta, May 14, 1942. This WPA crew is placing trees and laying cable anchor line on the severely eroded levee.

Tomé Hill, on the east side of the Rio Grande, is located on the Camino Real and is a National Historic Site. The sculptures, by Armando Alvarez, were erected in 1997.

skyline at 2:00–2:30, Cat Hills flow and cinder cones at 2:30, and Wind Mesa volcano at 3:00. The broad plain extending westward from the base of the Manzanita–Manzano range (8:00–11:00) is the Llano de Manzano. The north-trending break in slope from 8:00 to 10:00, about midway up the Llano, is the scarp of the Hubbell Springs fault. The scarp marks the eastern margin of the deep southern segment of the Albuquerque Basin. (0.4)

14.2 Crossing bridge over railroad. (0.8)

15.0 Wind Mesa volcano at 3:00 is a 4-million-year-old shield volcano. Shield volcanoes are formed by very fluid lavas that can flow for long distances. Wind Mesa volcano is composed mainly of basaltic andesite. (0.4)

15.4 Basalt flow in broad swale on 140-foot-high terrace surface to the right. This is the oldest of at least seven flows from the Cat Hills volcanic center and has been dated at about 100,000 years, making it one of the youngest volcanic features in the Albuquerque Basin. There are 23 aligned cinder cones in the Cat Hills field. Valencia County line just ahead. (1.3)

16.7 Tomé Hill is the 300-foot-high, dark hill across the river to our left. It is probably the remnant of a large volcano that has been severely eroded. The slopes are covered with colluvium, talus, eolian sand, and landslide blocks. Rocks from the core of the hill have been dated at 3.5 million years.

Tomé hill is known for the colorful Easter sunrise processions, as pilgrims ascend to the crosses and shrines on the summit. Petroglyphs on the hill indicate that its religious significance may stretch far back into prehistoric time. Recently the hill was placed on the National Register of Historic Places. El Camino Real extended past Tomé Hill, through Tomé and other small farming villages east of the Rio Grande.

In 1974 the Shell No. 1 Isleta Central test well, about 2 miles to the west, was drilled to a depth of 16,345 feet. The Santa Fe Group is 8,789 feet thick in this well, which bottomed in Permian strata. (2.3)

19.0 MP 204. Keep right for Los Lunas exit. The town, about 2 miles to the east, is named after the Luna family who settled here in 1808. The most famous building in town is the Luna Mansion, now a restaurant. When the Santa Fe Railroad surveyed a route through Los Lunas, the tracks went right through the original Luna family home. So, the railroad agreed to build the Luna family a new home constructed from adobe in the style of a southern mansion. (0.5)

19.5 Take Exit 203; prepare for right turn. (0.2)

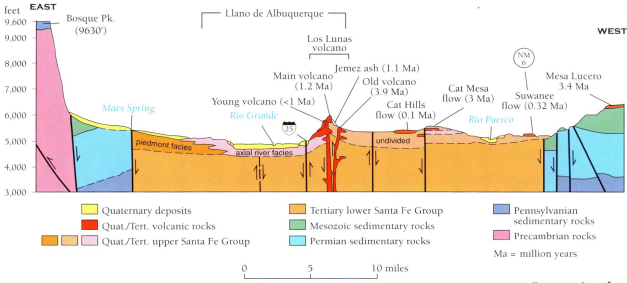

Cross section of the Albuquerque Basin near Los Lunas.

0.0 Stop light; reset trip meter. Turn right on NM–6. (1.0)

1.0 STOP 1. Pull off onto broad shoulder on the right. Directly across the road is the Los Lunas volcano (5,955 feet), which consists of at least two andesitic volcanoes. The older volcano, dated at 3.88 ± 0.04 million years, is an exhumed feature; it was partially buried by sands and gravels of the Santa Fe Group and later re-exposed by erosion. (The younger volcano was not buried by sediments.) Erosion has exposed a section of the older volcano along the southwest flank of the younger volcano, which has been dated at 1.22 ± 0.01 million years. Both volcanoes bent the sedimentary strata during eruption. The hummocky slopes on the northern flank of the volcano are landslides. The lower half of the slope at 10:00 is a landslide that uncovered a section of the northern feeder conduit of the Los Lunas volcanic center. Note the south-striking fault that has uplifted the lava flow and Santa Fe Group sediments over 100 feet between the northern conduit and the main volcano. (1.5) **Continue west on NM-6.**

2.5 Below the radio tower on the left are west-dipping upper Santa Fe Group strata. An erosion surface on the deformed Santa Fe beds is overlain by sandy alluvium and eolian deposits, with lenses of rounded gravel and scattered blocks of andesite derived from the Los Lunas volcano.

Los Lunas volcano.

Lenses of silicic volcanic ash are present at the base of this erosion-surface mantle. The ash is an air-fall unit from the last major Jemez eruption about 1.2 million years ago. The gray bench to the south is a 1-foot ash layer from the Los Lunas volcano. Sediments above the ash postdate the volcano. (1.3)

3.8 MP 29. Driving on alluvial apron of Los Lunas volcano. Mt. Taylor at 2:00, Cat Hills volcanoes at 2:30, and Wind Mesa volcano at 3:00. Llano surface is covered with bunch grass, yucca, scattered sage, and windblown sand. (0.8)

Unnamed cinder cone of the Cat Hills volcanoes.

4.6 Road left to Dalies on BNSF Railroad. About 3 miles to the south in 1952 the Long No. 1 Dalies test well was drilled to a total depth of 6,089 feet, all in Santa Fe Group sediments. (0.4)

5.0 Bridge over BNSF Railroad. Ahead on right are basalt flows 1 and 2 of the Cat Hills volcanic field. Flow 1 has been dated at about 100,000 years.

The predecessor to the AT&SF Railroad, the Atlantic and Pacific Railroad, began expanding west from Isleta in 1880, seeking to establish a second transcontinental route. By 1881 track was laid past the Arizona border, and in 1883 the railroad connected with the Southern Pacific Railroad at the Colorado River, completing the route. Once the long-planned Belen cutoff route through Abo Pass and on through eastern New Mexico was completed in 1907, traffic increased. The difficult grade at Raton Pass in northeastern New Mexico had made operations expensive, and the new, easier grade increased profits for the route.

Although little archaeological inventory has been conducted in this area, several prehistoric sites are known to exist along NM–6, most of them apparently farming sites that may have been inhabited only during the summer growing season by people from large pueblos some distance away. (0.8)

5.8 MP 27. Cross Duarte Road. Homes in this area must either carry in their own water or drill more than 500 feet to reach it. Socorro Mountains at 9:00, Magdalena Mountains at 9:30, Sierra Ladrones at 10:00, Bear Mountains at 11:00, and Lucero uplift from 11:30 to 1:00. (0.4)

6.2 Crossing a north-striking, west-down fault that cuts the Llano de Albuquerque surface. (1.6)

7.8 MP 25. Road curves to left. Broad swale ahead on Llano de Albuquerque surface is probably an erosional sag related to the Cat Hills–Wind Mesa zone of faults and fissures. (1.5)

9.3 Begin descent from Llano de Albuquerque into the Rio Puerco valley. Gravel and sand of Sierra Ladrones Formation (upper Santa Fe Group) are exposed in roadcuts and hillslopes on the right. (1.4)

10.7 Pottery Mound Pueblo site and University of New Mexico research area are on the valley floor to the south. This site is not open to the public. Mohinas Mountain is straight ahead. (1.1)

11.8 MP 21. Hidden Mountain straight ahead. Both Mohinas and Hidden Mountains are composed mainly of 8-million-year-old basalt and were exhumed after they had been buried by Santa Fe Group sediments. (1.3)

13.1 Hidden Mountain at 9:30, Lucero uplift (Mesa Lucero) at 10:00, and Mt. Taylor on the skyline at 11:30, behind basalt-capped Mesa Redonda in the middle distance. The lava on top of the Lucero uplift is 3.4 million years old. This uplift represents the eastern edge of the Colorado Plateau.

To the south is the famous Los Lunas Mystery Stone, a controversial petroglyph located near the base of Hidden Mountain. Several people have attempted a translation of the inscription on a basalt boulder. Early studies concluded that the text is the Ten Commandments, written in a combi-

Los Lunas Mystery Stone.

POTTERY MOUND

Pottery Mound, a large prehistoric pueblo scattered over 7 acres, lies near here on the west bank of the Rio Puerco. Inhabited from the early 1300s to 1475, the village consisted of three or four blocks of puddled adobe rooms that stood three or four stories high in places. The site takes its name from the multitude of pot sherds that once covered the ground. All 17 of the kivas were adorned with paintings applied to layer after layer of the adobe-plastered walls in depictions of religious ceremonies and figures of sacred importance in reds, oranges, pinks, salmons, vermilions, maroons, yellows, greens, blues, purples, lavenders, blacks, whites, grays, and (rarely) browns. Chemical analysis of the paints revealed that pigments were derived from iron oxides (hematite and limonite), cinnabar,

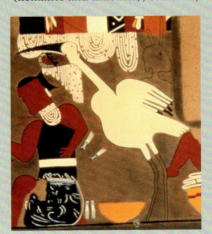

manganese, copper carbonates (malachite and azurite), powdered turquoise, gypsum, kaolin, chalk, and even uranium oxide. Many of the same minerals are used today as pigments in commercial paint manufacture. Several caches of minerals, some of them abraded, apparently from grinding on metates, were found in the kivas. Organic pigments such as charcoal and possibly coal were used for black colors. Binders and mediums were probably animal fats and oils, water, plant resins, saps, and juices, and perhaps the whites and yolks of bird eggs.

No attempt was made to save more than samples of the murals during the excavations of 1954–1962, but photographs and paintings were made as the murals were uncovered. Pottery Mound now belongs to the University of New Mexico.

Mohinas Mountain (left) and Hidden Mountain.

OPPOSITE:
The Rio Puerco in 1983. Note the narrow and vegetated floodplain.

BELOW:
Old highway bridge across the Rio Puerco in 1916. Note the wide, braided stream channel.

nation of several Near Eastern alphabets. Perhaps the most remarkable claim is that it is an account left by a lost Phoenician sailor about 500 B.C. Others suspect a more recent origin, possibly UNM anthropology students from the 1930s, which is the date on a similar carving nearby, written in English. Access to this area is restricted by Isleta Pueblo. (1.3)

14.4 STOP 2. Pull off into the parking area on the left before the bridge to view deeply incised channel of the Rio Puerco. Use caution here and watch for oncoming traffic.

About 5 miles south are the Gabaldon badlands, where the thickest section of the Santa Fe Group in the Albuquerque Basin is exposed. Within the 3,700 feet of west-tilted strata, many vertebrate fossils have been recovered, including camel, beaver, rabbit, fox, cat, horse, and pronghorn antelope. Access to this area is limited; most of it is on private land.

The northern part of the Lucero uplift, from 9:00–1:00, forms the eastern edge of the Colorado Plateau. The uplift is an asymmetrical anticline, plunging northward. Along this part of the Albuquerque Basin margin, steeply dipping Santa Fe Group beds are downfaulted and dragged against Permian and Triassic beds along the Santa Fe fault. The anticline transforms southward into the high-angle, reverse Comanche fault. Both features are probably Laramide in age, and formed before Late Oligocene time (approximately 30 million years ago). **Continue west on NM–6. (0.3)**

14.7 Entering Laguna Pueblo Indian Reservation and valley of Arroyo Garcia. The pueblo itself is 28 miles to the northwest. (1.8)

16.5 Ridge from 1:00 to 2:00 is an outlier of deformed Santa Fe Group deposits dipping northeast and capped with older

THE RIO PUERCO

The Rio Puerco drains much of northwestern New Mexico, with a drainage basin area of 6,590 square miles. A hydrologic gaging station near the railroad bridge to the south indicates an average maximum annual discharge of about 10,000 cubic feet per second. Notable floods occurred on August 12 and September 23, 1929. The flood of September 23 destroyed the railroad bridge, with an estimated discharge of 37,700 cubic feet per second.

The Rio Puerco has one of the highest suspended sediment loads of any river in the world, with an average of approximately 400,000 parts per million (ppm). The highest recorded sediment concentration for the Rio Puerco was about 680,000 ppm, 75 percent of which was sand: Imagine taking 10 gallons of this river water home in a bucket, letting the water evaporate, and finding 6.8 gallons of dirt in your bucket. Waters with such high concentrations of sediment are much heavier than pure water and, more capable of destroying bridges, carrying away automobiles, and silting up downstream reservoirs such as Elephant Butte.

The Rio Puerco has developed an active inner channel and inner floodplain within the Rio Puerco arroyo. Sediments within this channel commonly consist of sand and pebbly sand with armored mud balls. Point bars consist of gravelly and fine-grained sands; oxbows are commonly filled with laminated silt and clay.

The time of incision of the Rio Puerco is not well known. One interpretation is that the Rio Puerco became a continuously incised arroyo after 1885. There is documentary evidence of problems with gullying in this area as early as 1765, and there are descriptions of incised reaches (at least 20 feet deep) in 1846 and 1855. No early reports describe a shallow channel in this area. Rates of soil-pipe development, based on comparison of aerial photographs, suggest that the deeply incised arroyo is less than 200 years old.

At Pottery Mound, a similar-sized arroyo and tributary arroyo existed during occupation of the site. Trash was thrown into the arroyos as they filled with sediment. The old arroyos had filled with sediment, and the channel had moved northward before the current arroyo formed.

Known archaeological sites on the lower Rio Puerco arroyo show a pattern repeated in historic times. A few sites (such as Pottery Mound) are large and indicate stable settlements, but many are small, briefly occupied locations where people from Late Archaic times (about 500 B.C.) up through A.D. 1500 attempted hunting and plant collecting when drought struck and crops failed. One tiny camp a few miles south of NM–6 on a dry tributary of the Rio Puerco contains a handful of sherds from vessels made in the northern Mogollon area. The environment and character of the site suggest a farm camp inhabited for only one season by people in a desperate search for water associated with arable land.

Fossil camel tracks in the Sante Fe Group.

alluvium of Pliocene–Pleistocene age (Sierra Ladrones and younger units). The valley of Arroyo Garcia between this ridge and the Lucero uplift is part of a former, lower Rio San Jose valley, which was cut off by the Black Mesa–Suwanee basalt flow in middle to late Pleistocene time. (0.3)

16.8 MP 16. The light-colored deposits along the base of the Lucero uplift are travertine. (1.8)

18.6 Road ascends a low scarp at the southern tip of Black Mesa, part of the extensive Suwanee basalt flow described below. Contact of basalt flow on older sandy-gravel alluvium is exposed in the roadcut ahead. For the next 6 miles the road is on the Suwanee basalt flow, which lies 200 feet above the floor of the Rio Puerco valley. This flow originated from Cerro Verde volcano, about 14 miles to the southwest, and followed a 30-mile-long horseshoe path to this point around the northern end of the Lucero uplift via the ancestral valleys of Arroyo Lucero and lower Rio San Jose. The basalt flow is over 300,000 years old. (2.2)

20.8 MP 12. Note the Manzano and Sandia uplifts on the far eastern skyline beyond the Llano de Albuquerque. From this point you can see across the entire 30-mile width of the Rio Grande rift. (0.6)

21.4 Road curves to left. The structural boundary between the Colorado Plateau and the Rio Grande rift is beneath the basalt flow in this area. (2.4)

23.8 MP 9. STOP 3. Carefully pull off on right side of road.

On the west side of the road is Mesa Lucero, the northern end of the Lucero uplift. The mesa, with Lucero cone at its summit, is capped by 3.4-million-year-old

TRAVERTINE

Travertine is composed of calcium carbonate ($CaCO_3$) deposited as calcite and (less commonly) as aragonite, in a spring system. As carbonate-rich water is emitted from the spring, calcium carbonate is precipitated. Travertine deposits can build up as waterfall/cascades, as lake fills, or as mounds or ridges. Faults (such as those along the base of the Lucero uplift) are commonly channels for travertine springs. Although over 50 travertine deposits have been described in

New Mexico, only one produces travertine dimension-stone (quality building stone), the New Mexico Travertine quarry just south of here at Mesa Aparejo. The Mesa Aparejo stone is cut and polished at a mill near Belen. The New Mexico Travertine operation is one of only three in the U.S.; the others are in Idaho and Montana. Travertine is used as an architectural, decorative, or ornamental stone, for monuments, as patio floors, as flagstone, and as tiles.

Pliocene basalt and spring deposits overlying the Permian Yeso Formation, Glorieta Sandstone, San Andres Formation, and Triassic Moenkopi Formation and Chinle Group. The structural and stratigraphic relationships are obscured by many landslides on the mesa flanks. Sharp ridges along the base of the Lucero uplift are hogbacks formed of eastward-dipping Permian, Triassic, Jurassic, and Cretaceous sedimentary rocks.

The Lucero uplift marks the eastern rim of the Colorado Plateau; dips to the west within the uplift are only a few degrees. The eastern margin of the Lucero uplift is a narrow, structurally complex zone. The rift-forming Santa Fe fault lies only 0.5–1.0 miles to the west and drops upturned Santa Fe Group beds against Permian and Mesozoic beds along the Colorado Plateau margin.

From here northward along the Rio Puerco fault zone to the southern end of the Nacimiento uplift, a distance of about 45 miles, the rift border is not visible in the topography. The rift-filling Santa Fe Group beds (of the Llano de Albuquerque area) stand higher than the older, east-dipping

Mesa Lucero, on the west side of Highway 6 at Stop 3.

Cretaceous units of the uplifted border.

On the east side of the road (to the right) the view is of the canyon of the Rio San Jose, which joins the Rio Puerco about 3 miles to the southeast. Headwaters of this major tributary to the Rio Puerco are along the Continental Divide in the Zuni Mountains–Mount Taylor area, about 50 miles to the west. The canyon is about 140 feet deep, with another 100 feet of alluvial fill below the valley floor.

The canyon has eroded about 12 miles upstream from the Rio Puerco in 320,000 years, an average rate of 200 feet per 1,000 years. The sandstone cliff across Rio San

Canyon of the Rio San Jose.

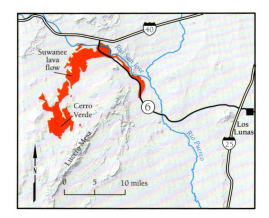

The Suwanee lava flowed north around the Lucero uplift and then south along the valley of the ancestral Rio San Jose.

Pressure ridge in the Suwanee basalt flow at mile 25.7.

Bluff on the far side of Rio San Jose at Stop 3.

Jose canyon is eolian Bluff and Zuni Sandstones in the rounded shoulders and the vertical face. A thin wedge of multi-colored mudstone, shale, and sandstone of the Morrison Formation (Jurassic) is above the Zuni and is capped by Cretaceous Dakota Sandstone with slight angular unconformity. Patches of Jurassic Todilto Formation and Summerville Formation are exposed below the Bluff Sandstone in the inner canyon. At the base of Mesa Redonda, the Todilto Formation (Tonque Arroyo Member) is overlain by the Summerville–Bluff–Zuni–Morrison–Dakota sequence. Mesa Redonda is capped by a small remnant basalt flow different from that of Mesa Lucero. The small cuestas and ridges between here and Mesa Redonda consist of faulted wedges of upper Morrison Formation, Dakota Sandstone, and Mancos Shale. **Continue on NM–6. (1.9)**

25.7 Mesa Redonda at 11:00; note the pressure ridge in the Suwanee basalt flow to right. The Entrada Sandstone and over-lying Todilto Formation are exposed in the low ridge ahead and in railroad cuts to the left. Note the Tonque Arroyo Member of

the Todilto Formation at the base of Mesa Redonda to left. Weathered gypsum under-lies much of the alluvial flat between here and the mesa base. The Tonque Arroyo gypsum was locally strip-mined during 1952 and 1954 by the Suwanee Gypsum Products Corporation and the White Eagle Gypsum Company. There are numerous uranium prospects in the basal Todilto Formation in this area. (1.4)

27.1 At 10:00, note faulted syncline in Bluff–Morrison–Dakota sequence exposed in north face of Mesa Redonda. This struc-ture is truncated by a basalt-capped sur-face along the mesa summit. (0.7)

27.8 MP 5. Site of Suwanee Station on BNSF Railroad to left. The original station name was San Jose but was changed to its present name in 1902. The upper end of the Rio San Jose canyon segment is to the right. The broad plain ahead is at the confluence of Arroyo Lucero from the south, Arroyo Colorado from the southwest, and Rio San Jose drainage systems. The valley is formed mainly on shale and mudstones of the Triassic Chinle Group. Locally, there is a thick alluvial and eolian cover of middle to late Quaternary age, and a discontinuous veneer of Pleistocene basalt (including the Suwanee flow). (0.5)

28.3 Mt. Taylor (elevation 11,389 feet) at 1:00. This peak, named after President Zachary Taylor, is a composite volcano consisting of rocks ranging in composition from andesite to rhyolite. Radiometric dates suggest that the volcano was active between 1 and 4 million years ago.

If you look on a regional geologic map, you will notice a northeast-trending string of young volcanoes consisting of, from southwest to northeast, the Springerville volcanic field of west-central New Mexico and east-central Arizona, the Malpais basalt field, the Mt. Taylor volcano, the Jemez Mountains caldera, the Taos Plateau volcanic field, and the Raton–Clayton and Ocate volcanic fields of northeast New Mexico. It has been suggested that this linear pattern of volcanism, called the Jemez lineament, exists because magma has leaked upward along a zone of weakness in the crust. A second less accepted theory suggests that as the tectonic plate (in this case the North American plate) moved over a stationary deep mantle upwelling or plume of anomalously hot magma (known as a hot spot), a line of volcanoes formed. The best-known examples of this phenomenon are the Hawaiian Islands and Emperor seamounts on the Pacific plate. However, if this is the case for the Jemez lineament, the volcanoes should be younger to the northeast; instead, all of the volcanic fields have similar ages.

For the Navajo, Mt. Taylor is the sacred mountain of the south, and is called DootL'ishiidziil (Navajo for "turquoise mountain"). Several Pueblo cosmologies also consider Mt. Taylor sacred. Navajo tradition holds that the mountain, home of Turquoise Boy and Yellow Corn Girl, is "fastened from the sky to the earth with a giant flint knife and decorated with

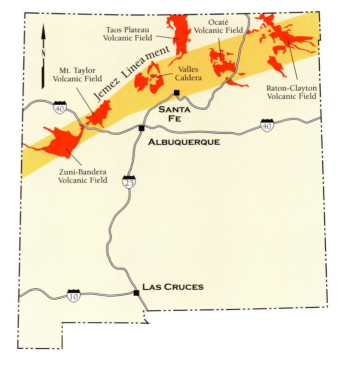

The Jemez Lineament is defined by a string of prominent northeast-trending young volcanoes.

turquoise, dark mist, female rain, and all species of animals and birds." The Mt. Taylor area is also the source of a distinctive pitchy black obsidian widely used for arrowheads and other tools during several prehistoric periods. (2.6)

30.9 Route crosses old US–66. Wild Horse Mesa Bar to right. Correo (Spanish for "mail") is the townsite at 9:00 across railroad overpass. The town had a U.S. Post Office from 1914–1959. It was the only spot in the area where mail could be received and dispatched, hence the town's name. Cibola County line. (0.7)

31.6 Crossing the Rio San Jose, which flooded on September 23, 1929, and August 21, 1935. Note abundant water-guzzling salt cedar along the channel. Cliff at 12:00 is composed of the Triassic Chinle Group. Here, sandstone overlies shale of Petrified Forest Formation. Farther west in Arizona, this unit contains petrified trees. Mesa Gigante is straight ahead. The excellent exposures in the cliffs to the north display strata from Triassic to Cretaceous. **Prepare to join I–40 east to Albuquerque just ahead.**

0.0 **Turn right onto I–40 east on-ramp. Reset trip meter.**

0.6 Low, rocky hills along the interstate on the left are landslide blocks that broke away from the Chinle Group cliff. (0.6)

1.2 MP 128. Suwanee Peak is at 2:00 and Mesa Redonda is at 2:30. Just ahead on the left, the gray and pink cliffs beyond the Triassic section are composed of Todilto Formation limestone overlying Entrada Sandstone. Note scattered juniper and piñon pine along both sides of the interstate. (1.1)

2.3 Bench and slopes at 9:00–10:00 are formed on Chinle Group along the southeastern base of Mesa Gigante, a large tableland. Low knobs at 2:00–2:30 are capped by tilted Dakota Sandstone over Morrison Formation. The westernmost fault of the Rio Puerco fault zone passes

Mt. Taylor is a composite volcano.

LAGUNA PUEBLO

The founding of Laguna Pueblo, unlike other New Mexico pueblos, did not occur until the return of the Spanish after the Pueblo Rebellion of 1680. Sometime after 1692 people from the pueblos of Cochiti, Cieneguilla, Santo Domingo, and Jemez, unhappy with the return of Spanish rule, left to join the people at Acoma and Zuni. In 1697 these 100 refugees, along with some Acoma residents, settled in what is now Old Laguna. The pueblo made peace with the Spanish in 1698 and received an official land grant. Eventually, the U.S. Government also recognized the grant. In time, the pueblo grew and now includes six villages and a large population. Once the holders of the largest sheep herds among the pueblos, most Laguna residents no longer farm or herd. Many were employed at the Jackpile uranium mine, until it closed in the late 1980s. A very successful Laguna-owned and operated defense contractor now employs many Laguna residents.

The languages spoken at Laguna and Acoma (and five other pueblos) belong to the Keresan family. Unfortunately, like many other American Indian languages, Keresan is being lost. Wick R. Miller, a linguist who studied the Acoma language, noted that in 1949 none of the children entering first grade spoke English, but by the late 1950s, several did. By 1960 nearly all adults spoke English, and only a few elderly people still spoke only Acoma.

northward through this gap between Chinle and Dakota outcrops. This fault zone marks the eastern boundary of the Colorado Plateau. Most of the north- to northwest-trending normal faults of the Rio Puerco fault zone are downthrown to the west. (1.1)

3.4 Lower Mesaverde Group–Gallup Sandstone is exposed in the roadcut on the left. Mancos Shale is exposed in gullies to the right. Low ridges (cuestas) from 1:30 to 3:00 are capped by basal Dakota Sandstone over Morrison Formation and Zuni Sandstone. Our route for next 4.5 miles crosses poorly exposed Cretaceous beds of sandstone and shale. (1.8)

5.2 MP 132. Entering Bernalillo County. (0.4)

5.6 Crest of ridge. The Sandia Mountains are on the skyline straight ahead. The Manzano Mountains at 1:00–2:00 form the eastern border of the Rio Grande rift. The high tableland in middle distance is the Llano de Albuquerque. (0.6)

6.2 Crossing area with thin cover of older alluvium on Mesaverde Group sandstones and shales. (1.2)

7.4 Route crosses West Apache fault and western boundary of the Apache graben. Basin fill in the graben is the westernmost occurrence of Santa Fe Group in the Rio Puerco fault zone. (2.7)

10.1 Bridge over Cañada de los Apaches, a tributary of the Rio Puerco. (1.1)

11.2 MP 138. Crossing the eastern part of Apache graben. Deformed pebbly sand and clayey silt beds of the upper Santa Fe Group are visible in the roadcut to the right. (0.6)

11.8 Thick, white lens of volcanic ash in ancestral Rio Puerco terrace caps the small knob to the left. This ash, used for sand-casting silver buckles and bracelets, is from a 600,000-year-old eruption in what is now Yellowstone National Park. Our route crosses East Apache fault and eastern

RIO PUERCO ARCHAEOLOGY

As barren and desolate as the Rio Puerco is now, one would be justified in believing it has always been thus—not so, as dozens of archaeological sites attest. Scattered over the floodplain both north and south of I–40 in a stretch just 3.75 miles wide by 9 miles long are 116 archaeological sites representing 141 occupations that date from the Middle Archaic (about 3200 B.C.) up to the latter part of the 19th century. The most intense occupation began in Basketmaker III times (A.D. 500–700), just as Anasazi people began farming in earnest, and continued through Pueblo III times (A.D. 1300). A few later sites in the Pueblo IV period (A.D. 1300–1600) were probably left by people from big pueblos on the San Jose to the west or the Rio Grande to the east—people who came to the Rio Puerco only in the summer to plant and tend their fields of corn, beans, and squash. Many sites are small (a quarter of an acre or less), but some are as large as 20 acres.

In the 1750s several widely scattered Hispanic farming villages were settled along the Rio Puerco from near its headwaters at Cuba, 70 miles to the north, to the confluence of the Rio Puerco and Rio San Jose, 12 miles to the southwest. However, all the villages were deserted by 1774 because of weather conditions and

Chaco Black-on-white jar, A.D. 1050-1200.

Navajo raids. New Hispanic settlers, some descendants of the first settlers, established small farms and villages in the 1870s and tried to raise livestock and crops. One of the Rio Puerco's downcutting cycles (which have been repeated over tens of thousands of years) accelerated in the late 1880s, coinciding with a river downcutting cycle over much of the western U.S. Probably exacerbated by overgrazing, the rapid entrenchment caused the river to drop to the point that by 1900 irrigation water was unobtainable at the lower villages. At San Luis on the upper river, the last dam washed out in 1951. Except for a smattering of optimistic ranchers who still run a few head of cattle, the upper Rio Puerco is deserted because of the lack of water. The archaeological sites, however, prove that, if never tropical, the Puerco was once green.

edge of Apache graben. Upper Santa Fe Group sediments are downthrown to the west against eastward-dipping mudstones and yellowish-brown sandstones of Late Cretaceous age, just ahead on the left. The Shell No. 1 Laguna Wilson well just south of the freeway started in Upper Cretaceous strata and encountered Precambrian rocks at 11,102 feet. (1.2)

13.0 Outcrop of sandstone and mudstone to the left is the easternmost exposure of Upper Cretaceous bedrock in the Rio Puerco fault zone in this area. Between here and the Llano de Albuquerque is the western margin of the Rio Grande rift. La Mesita Negra is the small, basalt-capped hill just east of the Rio Puerco Trading Post at 11:30. Santa Fe Group basin fill thickens markedly to the east of this area. (0.2)

13.2 MP 140 and exit to Rio Puerco–Cañoncito area. (0.5)

13.7 Crossing deeply incised inner channel of the Rio Puerco. This deep channel became established on the west side of the Rio Puerco valley between 1888 and 1896, apparently as a result of dam failures and dam diversions upstream. The pre-1888 channel remains a swale at the base of La Mesita Negra. (0.3)

14.0 Basalt-capped La Mesita Negra, dated at 8.11 ± 0.05 million years, is to the left. Tilted middle Santa Fe Group red beds are exposed below basalt on the north face of the mesita. Route ahead ascends escarpment to the Llano de Albuquerque. (1.4)

15.4 Cerro Colorado at 2:30 is a complex volcanic center emplaced within the Santa Fe Group. No age has been determined for this feature, but it is probably at least several million years old. (1.6)

17.0 Fine-grained, pinkish sediments in ridges and roadcuts to the left are lower Santa Fe Group. (0.2)

17.2 MP 144. Roadcuts ahead are in gravel and sand of the upper Santa Fe Group. (1.0)

18.2 MP 145. Route crosses the valley rim onto the Llano de Albuquerque. The rim is marked by a white layer of pedogenic (formed as part of soil-making process) calcium carbonate. A thin cover of eolian sand overlies the surface. To the north and south, the rim is locally capped by large sand dunes. For the next 4.5 miles the route crosses the Llano de Albuquerque, a former central-basin alluvial plain that is now preserved as an extensive tableland summit. Aggradation ceased after initial river-valley incision by the Rio Grande and Rio Puerco about 1 million years ago. The route across the

Cerro Colorado, visible at mile 15.4, is an exhumed volcanic center that was emplaced in the Santa Fe Group.

As Interstate 40 drops off the Colorado Plateau into the Rio Grande valley, a remarkable view of the city of Albuquerque, nestled between the Rio Grande and the Sandia Mountains, presents itself.

Llano de Albuquerque passes through a partially stabilized late Quaternary dune field composed of greatly elongated northeast-trending ridges of sand. (1.9)

20.1 In 1948 the Carpenter Atrisco No. 1 well just north of the freeway was drilled to a depth of 6,654 feet. The dry well penetrated 3,300 feet of Santa Fe Group sediments and bottomed in lower Tertiary strata. (1.6)

21.7 Albuquerque volcanoes are at 9:30, Jemez Mountains are directly behind them in the far distance. Sangre de Cristo Mountains at 10:30. The string of small volcanoes and associated spatter cones of the Albuquerque volcanoes are aligned along a north-trending fault system that includes the County Dump fault to the south. The oldest flow from the center is nearly 200,000 years old. See Trip 5 for a tour of the volcanoes. (1.4)

23.1 We now cross the rim of Cejita Blanca scarp and begin our descent into the Rio Grande valley. Note thick soil in roadcut to right. Cuts and outcrops of upper Santa Fe gravel and sand to left are on the upthrown block of the County Dump fault.

This descent marks the boundary between the Colorado Plateau and the Rio Grande rift. Most people consider the Rio Grande rift a northern arm of the Basin and Range province. Whatever the case, the abrupt change in landscape is a reflection of some fundamental differences in geologic structure between the two provinces. (0.3)

23.4 The County Dump fault is exposed in the roadcut on the left. Surficial sediments and buried soils of the downthrown blocks are exposed in the roadcut on the right. This fault extends northward and served as the conduit for lavas that formed the Albuquerque volcanoes. On November 28, 1970, a Richter magnitude 3.2 earthquake was felt throughout Albuquerque. The next year on January 4, a second earthquake of magnitude 3.5 rumbled through Albuquerque, causing about $40,000 damage to the University of Albuquerque. Both earthquakes are believed to have occurred along the County Dump fault zone, although no evidence of surface rupture was reported.

Just ahead is a remarkable overview of Albuquerque and the Sandia Mountains, as the highway continues its descent into the Rio Grande valley. The city is located within an inner valley cut by the Rio Grande during the last million years. The surface elevation of the river is about 4,800 feet; the higher peaks of the Sandia Mountains are over 10,000 feet. Two hundred years ago much of the Rio Grande valley was bosque forest. Farming, flood control, urbanization, and wildfires have reduced it to a ribbon of green between Cochiti Dam to the north and Elephant Butte Lake to the south. (3.8)

27.2 MP 154. Route descends to Segundo Alto surface; here an extensive fill terrace is on the Los Duranes alluvium. This surface represents a Rio Grande level during late Pleistocene time. (1.2)

28.4 Coors Road underpass. Route descends to valley floor from Segundo Alto surface. Roadcuts ahead are in sand and clayey silt of the Los Duranes alluvium. (0.7)

29.1 Western abutment of Rio Grande bridge. Route crosses modern conveyance channel and floodway. (1.1)

30.2 Exit to Rio Grande Boulevard. Historic Old Town and museums are to the right.

Albuquerque Museums

Not surprisingly, Albuquerque area museums reflect the area's history and environment. The following museums (by no means a comprehensive list) offer remarkable collections, as well as opportunities to learn and see more concerning the geologic and cultural history of the region. Phone numbers (and even locations) sometimes change; check the local phone book before heading out.

ON THE UNIVERSITY OF NEW MEXICO CAMPUS

Geology Museum, Department of Earth and Planetary Sciences
Northrup Hall
(505) 277-4201
epswww.unm.edu/museum.htm

Maxwell Museum of Anthropology
(505) 277-4405
www.unm.edu/~maxwell

University of New Mexico Art Museum
(505) 277-4001
unmartmuseum.unm.edu

Jonson Gallery of the UNM Art Museum
1090 Las Lomas NE
(505) 277-3188
www.unm.edu/~jonson

OTHER NOTEWORTHY MUSEUMS

Albuquerque Biological Park
(505) 764-6200
www.cabq.gov/biopark
Includes the Aquarium and the Rio Grande Botanic Garden (at 2601 Central Avenue NE), and the Rio Grande Zoo (903 10th Street SW).

The Albuquerque Museum
200 Mountain Road SW
(505) 243-7255
www.cabp.gov/museum/index.html

Casa San Ysidro
Corrales
(505) 898-3915
www.cabp.gov/museum/education/casa.tour.html
This beautifully restored Spanish-Colonial house is operated by the Albuquerque Museum; tours are available by appointment only.

Coronado State Monument
Bernalillo, on NM–44 west of I-25
(505) 867-5351
The murals from Kuaua, some of the very few prehistoric painted murals in

the Southwest, are beautifully displayed in the museum here.

Indian Pueblo Cultural Center
2401 12th Street NW
(505) 843-7270
www.indianpueblo.org

National Atomic Museum
1905 Mountain Road NW (next door to NMMNHS)
(505) 245-2137
www.atomicmuseum.com

National Hispanic Cultural Center of New Mexico
1701 4th Street SW
(505) 246-2261
www.nhcnm.org
This museum opened its doors in December 2000. In addition to exhibits, the center includes the Cervantes Institute for teaching Spanish and other classes, a library focused on history and genealogy, gift shop, and restaurant.

New Mexico Museum of Natural History and Science
1801 Mountain Road NW
(505) 841-2800
www.museums.state.nm.us/nmmnh
Includes the Lodestar Astronomy Center, the Dynatheater, and exhibits on dinosaurs, volcanoes, geology, and biology. A new museum, the Explora Science Center and Children's Museum, is slated to open in 2003 just east of the museum. The new museum will stress the integration of art, science, culture, and technology, and will feature interactive exhibits appropriate to all ages.

Turquoise Museum
2107 Central Avenue NE
(505) 247-8650
Located near Old Town, the museum includes turquoise from over 50 mines.

Glossary

acequia a cooperatively owned and maintained irrigation ditch. Such ditches are widely used in the northern Rio Grande valley.

aggrade to build up sediments within the channel of a stream in order to maintain its grade.

alluvium unconsolidated mud, sand, and gravel deposited by rivers, streams, or sheetwash.

alluvial fan a fan-shaped mass of sediment deposited by a stream at the mouth of a canyon in arid or semiarid regions.

ammonite an extinct single-shelled animal, of the cephalopod class, related to the present-day chambered nautilus. Found in Jurassic-to-Cretaceous rocks.

anticline an arch-like fold in which strata dip in opposite directions from a common ridge or axis; the core of an anticline contains the older strata.

aquifer a body of rock or sediment that is capable of storing and transmitting water freely.

basalt a dark-colored, generally volcanic rock, rich in iron and magnesium, which typically erupts from fissures and low-relief volcanoes.

brachiopod a bivalved marine invertebrate, commonly found as fossils in Paleozoic rocks of this region. Superficially resembles a clam, but each valve or shell of a brachiopod is bilaterally symmetrical. Range is lower Cambrian to present.

bryozoan a small, colonial aquatic animal, usually secreting calcareous skeletons that, as fossils, resemble either twigs or moss-like patterns. Range is Cambro-Ordovician to present.

caldera a large volcanic crater, often 10-20 miles in diameter, that forms when a gas-charged viscous magma body explosively erupts. Following the eruption, the shallow roof of the chamber collapses, forming a large, circular depression.

cephalopod a marine mollusk with well-defined head, eyes, and tentacles around the mouth; modern cephalopods include squid, octopus, and cuttlefish. Range is Cambrian to present.

colluvium loose fragments of rock and soil that accumulate at the base of slopes.

concretion a rounded body found in sedimentary rocks; usually caused by chemical deposition of concentric layers of calcite or silica around a central nucleus. Geodes are one type of concretion.

contact the surface between two types, ages, or units of rock.

correlative belonging to the same age or stratigraphic position.

crinoid an echinoderm with numerous radiating arms, typically (but not always) attached to the sea water. The segmented stems of Paleozoic crinoids are especially abundant as fossils in Mississippian and Pennsylvanian rocks. Range is from Ordovician to present.

crossbedding a type of bedding, typically found in sand dunes and stream-laid deposits, in which inclined internal layers are deposited at an angle to the predominant bedding orientation. Crossbedding indicates currents of moving water (or air) and can often be used to determine the direction from which these currents came.

cuesta a hill or ridge with a gentle slope on one side and a steep slope on the other.

eolian deposited by wind.

fanglomerate a sedimentary rock that is composed of rock fragments that were deposited in an alluvial fan.

fault a fracture along which movement has occurred.

fluvial deposit a sedimentary deposit transported and laid down by a stream.

gastropod a mollusk characterized by a distinct head with eyes and tentacles, and, in most cases, a single, unchambered shell of calcite. Modern-day gastropods include snails and abalone. Range is upper Cambrian to present.

greenstone a dark-green, metamorphic rock that forms through the metamorphism of certain mafic igneous rocks (such as basalt). Chlorite is a common mineral in greenstone.

horst an elongate uplifted block that is bounded on either side by faults.

igneous a rock that has formed from molten material. Igneous rocks can crystallize at or near the surface (as in extrusive, volcanic rocks like basalt) or deep beneath the surface (as in intrusive rocks, like granite).

inoceramid a large bivalved mollusk (clam) found in rocks of Cretaceous age.

laccolith a domed igneous intrusion, generally with a flat floor, which is concordant with surrounding rocks.

lava molten rock (magma) that erupts at the surface and hardens into a volcanic or extrusive igneous rock such as basalt.

limestone a sedimentary rock consisting chiefly of calcium carbonate (calcite). Limestones can form by either organic or inorganic means.

lode a mineral deposit in the form of an unusually large vein or a zone of veins in consolidated rock.

lycopod a giant club moss that grew to the size of a tree; fossilized remains are found in rocks of Mississippian and Pennsylvanian age.

maar a low-relief volcanic crater formed from numerous shallow steam and magma explosions.

mafic referring to (typically dark-colored) igneous rocks that are rich in iron and magnesium.

magma molten rock beneath the earth's surface. Once it erupts onto the earth's surface it is known as lava.

metamorphic rock a rock that has formed from pre-existing rocks at depth, without melting, in response to changes in heat, pressure, stress, and the chemical environment.

orogeny mountain building episode, generally involving large-scale faulting, magma intrusions, volcanic activity, and folding of rock layers in the earth's crust.

pelecypod a bivalved, bottom-dwelling aquatic mollusk that displays bilateral symmetry and a hatchet-shaped foot. Modern examples include clams and mussels. Range is from Ordovician to present.

piedmont a flat to gently sloping surface formed at the base of a mountain. In the Albuquerque area, piedmonts are thick alluvial aprons (in other parts of the country they may be cut into bedrock and covered with a thin veneer of alluvium).

placer mineral deposit formed by mechanical concentration of heavy mineral particles, such as gold, from weathered debris.

porosity the percentage of a rock or deposit that consists of open spaces.

porphyry an igneous rock containing two distinctly different size mineral crystals, typically large crystals (phenocrysts) in a fine-grained groundmass. The texture of such a rock is said to be porphyritic.

rift a long, narrow trough that is bounded on either side by normal faults.

sauropod a large four-footed, plant-eating dinosaur with a long neck and tail.

sedimentary rock a rock that results from the consolidation of loose sediment that has accumulated in layers, either clastic sediments (as in the case of sandstone) or chemically/organically precipitated sediments (as in the case of limestone).

strata layers of rock, usually sedimentary but may include volcanic flows.

syncline a trough-like fold in layered rocks; the core of a syncline contains the youngest rocks.

talus a sloping heap of rock fragments at the foot of a cliff or steep slope.

terrace any long, narrow, gently sloping surface bounded along one side by a steeper descending slope, and along the other by a steeper ascending slope.

Commonly found along the margin and above the level of some body of water, denoting a former level of the water. An inset stream terrace is a terrace that formed within a stream valley; it represents a former valley floor.

trilobite an extinct class of marine animals having a flattened segmented body covered by a dorsal exoskeleton divided into three lobes. Belongs to the phylum Arthropoda, which includes crustaceans such as lobsters and crabs. Fossil remains are found exclusively in rocks of Paleozoic age.

unconformity a surface that separates older and younger rock units and represents a gap in the geologic record. It may represent a time during which deposition was interrupted by a period of erosion, or a time of no deposition.

vesicular refers to the texture of a lava characterized by an abundance of vesicles or cavities, formed by the expansion and entrapment of gas bubbles that occurs as molten lava approaches the surface.

A Word About Maps

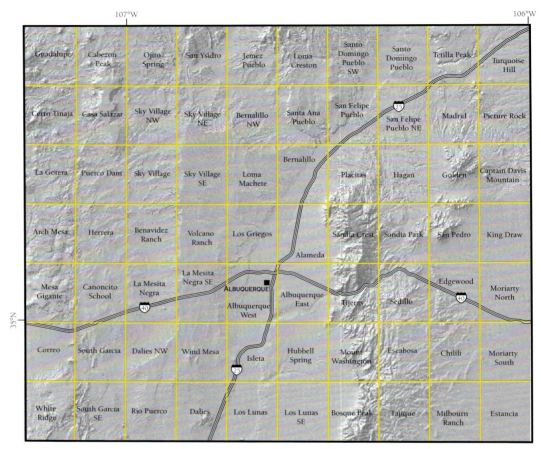

The index map above shows the distribution of 7 1/2-minute quadrangles across the Albuquerque area. These maps and a wide variety of other maps and publications are available from the New Mexico Bureau of Geology and Mineral Resources in Socorro. The bureau carries a complete inventory of U. S. Geological Survey maps for the state of New Mexico. Visit our Publication Sales Office on the campus of New Mexico Tech in Socorro, phone us at (505) 835-5410, or visit our Web site at www.geoinfo.nmt.edu

A wide selection of map products are also available locally in the Albuquerque area at:

Holman's
6201 Jefferson NE
Albuquerque, NM 87109
(505) 343-0007

Recreational Equipment Incorporated (REI)
1550 Mercantile Avenue NE
Albuquerque, NM 87107
(505) 247-1191

Index

A

Abert, J. W. 25
Abo Formation 20, 35, 85, 87, 88, 96, 102
Acoma Pueblo 57, 159
Alameda 65, 125, 132
Alameda Pueblo 132
Albert G. Simms City Park 68, 76-77
Albuquerque Basin v, 5-6, 7, 22, 23, 30, 31, 36, 39, 40, 41, 70, 75, 81, 92, 99, 117, 126, 142, 143, 147, 148, 149, 152
Albuquerque Metropolitan Arroyo Flood Control 75, 76
Alcanfor 56, 126
Algodones 62, 117
Alvarado Hotel 63
Anasazi 52, 77, 160
Ancestral Rocky Mountains 19
Apache Indians 26, 57, 60, 81
Arroyo Colorado 157
Arroyo de la Barranca 126
Arroyo de las Calabacillas 124
Arroyo Garcia 152, 154
Arroyo Lucero 154, 157
Arroyo Peñasco Group 18
Atchison, Topeka, and Santa Fe Railroad 62, 63, 64, 106, 110, 111, 129, 148, 150
Atrisco 43, 65, 162

B

Baca Canyon 72
Balsam Glade 90, 91, 93
Bandelier Tuff 106
Barelas Bridge 2, 62
Basketmaker II 51, 53
Basketmaker III 52, 53, 130, 160
Bear Canyon 68, 77
Bear Canyon Pueblo 77
Bear Mountains 150
Belen 22, 28, 150, 154
Bernalillo 39, 49, 62, 99, 118, 119, 127, 128, 129
Bernalillo County 131, 133, 159
Bigelow, John 26
Black Butte 22
Black Mesa 146, 154
Bluff Sandstone 156
Boca Negra Canyon 122
Boca Negra Cave 122
Bosque Redondo 57, 61
Bryan, Kirk 26
Budaghers 116
Burlington Northern Santa Fe Railroad 110, 115, 128, 150, 157. See also Atchison, Topeka, and Santa Fe Railroad
Bursum Formation 19

C

Cabezon Peak 74, 97, 98, 106
Cañada de los Apaches 160
Canjilon Hill 118
Cañoncito Navajo Reservation 57
Capulin Canyon 93
Capulin Spring Picnic Area 91
Carlito Spring 84
Carnuel 81, 85
Carson, Kit 57, 61
Carthage–La Joya Basin 22
Casa San Ysidro 125
Cat Hills 24, 31, 148, 150
Cebolleta 57
Cebolleta Mesa 57
Cedar Crest 84, 86
Cedro 81
Ceja Member 23
Cejita Blanca 162
Cerrillos 22, 25, 33, 35, 36, 87, 100, 101, 104, 107, 108, 109, 110

Cerrillos Hills 33, 102, 107, 110, 111, 112
Cerro Colorado 161
Cerro Columbo 103
Cerro Verde volcano 154
Cerros del Rio 112
Chaco Canyon 53, 101
Chinle Group 20, 87, 103, 105, 155, 157, 158
Cibola granite 81
Cibola National Forest 70, 88, 94
Civilian Conservation Corps 88, 90
coal 12, 19, 21, 31, 36, 96, 101, 106, 108, 111, 151
Cochiti Lake 47, 120, 128, 163
Cochiti Pueblo 60, 114, 116, 159
Colorado Plateau 7, 106, 142, 151, 152, 154, 155, 159, 162
Continental Divide 155
Coronado State Monument 55, 56, 119, 127-128
Coronado, Francisco Vásquez de 56-58, 69, 126
Corrales 55, 119, 124, 125
County Dump fault 122, 135, 137, 162, 163
Cunningham Hill mine 35

D

Dakota Sandstone 11, 21, 35, 96, 97, 106, 156, 159
Darton, Nelson H. 26
Desert Archaic 51-52, 130
Devil's Throne 110, 111
Diné 56, 57. See also Navajo Indians, Apache Indians
Doc Long Picnic Area 89, 90
Dry Camp Picnic Area 91
Dutton, Clarence E. 26

E

earthquakes 12, 14, 27-31, 135, 163

Edgewood 103
Edison, Thomas 104, 108, 109
El Camino Real 62, 128, 131, 148
Elena Gallegos Grant 76, 131
Elena Gallegos Picnic Area 73, 76
Ellis Trailhead Scenic Byway 92
Embudo Arroyo 77
Embudo Canyon 77
Entrada Sandstone 20-21, 35, 87, 105, 156, 158
Española Basin v, 114
Espinaso Formation 114, 116
Espinaso Ridge 116
Estancia Valley 5, 62

F

Falconer, Thomas 25
Flood Control Act 115
Fort Wingate 57
Four Hills 78, 80

G

Gadsden Purchase 45
Galisteo Basin 55, 109
Galisteo Creek 108, 110, 111, 113, 115
Galisteo Dam 115
Galisteo Formation 22, 114, 115, 116
Galisteo River 114
Galisteo–El Rito Basin 22
Gallup 36, 57
Gallup Sandstone 159
Glorieta Sandstone 20, 155
gold 25, 32, 33, 34, 35, 56, 58, 101, 103, 104, 105, 107, 108, 109
Gold Fields Mining Corporation 35, 103, 104

H

Hagan Basin 36, 93
Hidden Mountain 23, 31, 151, 152
Hubbell Bench 6

I

Isleta diversion dam 147
Isleta Pueblo 51, 55, 56, 60, 62, 64, 78, 132, 142, 145, 146, 147, 150, 152
Isleta volcano 23, 31, 144, 145, 146, 147

J

Jackpile uranium mine 159
Jemez Dam 128
Jemez lineament 157
Jemez Mountains 92, 96, 98, 106, 112, 113, 117, 126, 130, 137, 139, 141, 143, 157, 162
Jemez Pueblo 60, 159
Jemez River 117, 128
Juan Tabo Recreation area 33, 68, 70, 99
Juan Tomás 81

K

Kelley, Vincent C. v, 39, 40
Kinney Brick Quarry 84
Kuaua 55, 56, 126, 127, 128

L

La Bajada fault 110, 112, 113, 114, 115
La Bajada Hill 62, 113
La Cueva Canyon 33
La Luz Trail 33, 70, 93
La Mesita Negra 161
Ladron Mountains. See Sierra Ladrones
Ladron Peak 99, 139
Laguna Pueblo 57, 60, 62, 65, 142, 152, 159
Laramide orogeny 21, 22, 84, 96, 98, 152
Las Huertas 96
Las Huertas Canyon 94
Las Huertas Creek 94, 95, 117
Las Huertas fault 33, 96
Las Huertas Picnic Ground 94
Lee, Willis T. 42

Llano de Albuquerque 5, 24, 98, 124, 135, 141, 142, 150, 154, 155, 159, 161, 162
Llano de Manzano 148
Llano de Sandia 68, 124
Los Candelarias 131
Los Cerrillos. See Cerrillos
Los Duranes 64, 131
Los Gallegos 131
Los Griegos 64, 131, 132
Los Lunas 142, 148
Los Lunas Mystery Stone 151
Los Lunas volcano 23, 31, 142, 144, 147, 149, 150
Los Poblanos 131
Los Ranchos de Albuquerque 130, 131
Lucero uplift 7, 23, 35, 150, 151, 152, 154, 155, 156

M

Madera Group 11, 19, 35, 36, 46, 83, 84, 85, 87, 88, 89, 90, 91, 92, 93, 94, 96, 102, 103, 105
Madrid 87, 100, 101, 104, 106, 108, 111
Magdalena Mountains 74, 139, 144, 150
Malpais basalt field 157
Mancos Shale 21, 36, 85, 97, 98, 105, 106, 110, 111, 112, 114, 156, 159
Manzanita Mountains 6, 28, 32, 80, 81, 84, 122, 144, 148
Manzano Base 80
Manzano Mountains 6, 7, 22, 78, 81, 85, 92, 122, 148, 154, 159
Marcou, Jules 24, 26
Martineztown 45, 64
Menefee Formation 112
Mesa del Sol 144, 147
Mesa Gigante 158
Mesa Lucero 147, 151, 154, 155, 156
Mesa Verde 54, 55

Mesaverde Group 21, 35, 36, 85, 106, 108, 159
Mesita de Juana Lopez, 113
Mexican rule 45, 61
Middle Rio Grande Conservancy District 47-49, 120, 145
Modified Mercalli intensity scale 27, 28
Moenkopi Formation 20, 155
Mohinas Mountain 23, 31, 151, 152
Monte Largo 91, 102, 117
Montezuma Ridge 96
Morrison Formation 17, 21, 71, 86, 87, 96, 97, 98, 105, 114, 156, 158, 159
Mt. Taylor 26, 74, 92, 99, 137, 139, 144, 150, 151, 155, 157, 158
Mt. Chalchihuitl 33, 107, 109, 110

N

Nacimiento Mountains 98, 106, 137, 139, 141
Navajo Indians 52, 57, 60, 61, 65, 157, 160. See also Diné
New Mexico Museum of Natural History and Science 99
New Placers mining district 34, 101, 105
Nine Mile Picnic Area 91
North American Cordillera 10
North Valley 64, 76, 131
Northeast Heights 43, 62, 78

O

Old Placers mining district 32, 35
Oñate, Don Juan de 44, 58
Oro Quay Peak 103
Ortiz Mine 103, 104, 105
Ortiz Mountains 25, 32, 34, 35, 87, 93, 101, 103, 104, 105, 108, 110, 111, 117

P

Paa-ko 101
Paleo-Indian 51, 63, 79, 94, 95
Palomas Peak 93
Pecos Pueblo 55, 59
Pedernal Highlands 19
Petroglyph National Monument 31, 55, 119, 121-123, 135, 137
Picuris Pueblo 56, 60
Pike, Zebulon M. 25
Pino Canyon 73
Pino trail 76
Placitas 17, 33, 36, 42, 46, 79, 81, 93, 94, 96, 97
plate tectonics 10, 12-15, 17, 22
Popotosa Formation 22
Pottery Mound 56, 127, 142, 151, 153
Precambrian Era 17
Primera Agua 81
Pueblo I 52-53, 130
Pueblo II 53, 55
Pueblo III 53-55, 94, 160
Pueblo IV 55-56, 122, 160
Pueblo Rebellion 59-61, 109, 132, 159

R

Ranchitos 81, 117
Raton 36, 150, 157
Redonda Mesa 151, 156, 158
Redondo Peak 106
Richter scale 27, 163
Rincon Ridge 17, 69, 99, 117, 128
Rio Bravo del Norte. See Rio Grande
Rio Grande 2, 5, 7, 26, 36, 39, 40, 41, 45, 47, 49, 53, 56, 59, 61, 62, 78, 87, 102, 105, 106, 117, 119, 121, 122, 124, 128, 129, 130, 132, 138, 141, 142, 144, 145, 147, 160, 161, 163
Rio Grande basin 42, 46, 49, 120, 147
Rio Grande Nature Center 45, 119, 133-134

Rio Grande rift v, 5, 6, 7, 14, 22-24, 26, 27, 28, 29, 31, 84, 92, 96, 98, 106, 112, 114, 138, 154, 159, 161, 162
Rio Grande valley vi, 1, 5, 25, 28, 31, 42, 43, 48, 99, 111, 117, 143, 162
Rio Grande Water Compact 47
Rio Puerco 5, 7, 23, 25, 26, 36, 57, 62, 65, 125, 142, 143, 146, 151, 152, 153, 154, 155, 158, 160, 161
Rio Rancho 119, 125, 126
Rio San Jose 23, 154, 155, 156, 157, 158, 160
Rio San Jose canyon 155-156, 157
Route 66. See U.S. Route 66

S

San Andres Formation 20, 105, 155
San Antonio 81, 85
San Antonio Arroyo 86, 122
San Antonio Creek 81
San Antonio de las Huertas 96
San Antonio de Padua 81
San Felipe Pueblo 60, 62, 87, 102, 113, 116, 117
San Juan Basin 36, 53, 56
San Juan Pueblo 60
San Juan River 130
San Marcos Arroyo 110
San Marcos Pueblo 109
San Pedro 34, 81, 100, 101, 103
San Pedro Creek 87, 101, 102
San Pedro mine 34, 103
San Pedro Mountains 25, 34, 87, 91, 101, 103, 105
San Pedro Spring 102
San Pedro Valley 91, 93
San Ysidro 21, 86, 118, 125
Sandia Base 64, 78, 80, 146
Sandia Cave 51, 79, 94
Sandia Crest 5, 7, 33, 68, 70, 71-73, 79, 80, 87, 92, 93, 102, 124

Sandia Crest National Scenic Byway 90
Sandia Formation 11, 18, 19, 35, 88, 93, 130
Sandia granite 17, 33, 69, 70, 71, 77, 80, 81, 83, 88, 89, 93, 99
Sandia Mountain Wilderness 72, 76, 88
Sandia Mountains v, vi, 1, 2, 5, 7, 9, 11, 17, 20, 22, 31, 32, 35, 43, 65, 68, 69, 75, 76, 78, 79, 80, 81, 85, 86, 87, 88, 92, 97, 105, 106, 112, 117, 122, 124, 126, 128, 129, 137, 141, 144, 159, 162, 163
Sandia Peak Ski Area 73, 90, 91
Sandia Peak Tramway 33, 68, 71, 72
Sandia Pueblo 49, 53, 56, 60, 64, 65, 68, 69, 78, 99, 119, 128
Sandia Pueblo archaeological sites 128, 130
Sandoval County 99, 102, 116, 128
Sangre de Cristo Mountains 19, 92, 93, 101, 108, 112, 126, 137, 162
Santa Ana Mesa 49, 98, 117, 126
Santa Ana Pueblo 60, 116, 117
Santa Barbara 64
Santa Domingo 159
Santa Fe 20, 28, 33, 47, 55, 59, 61, 62, 63, 69, 85, 87, 92, 93, 108, 113, 120, 146
Santa Fe County 102
Santa Fe Group 6, 22, 23, 36, 39, 43, 80, 97, 99, 103, 114, 115, 116, 118, 121, 126, 128, 137, 141, 142, 143, 144, 145, 146, 147, 148, 149, 150, 151, 152, 155, 160, 161, 162
Santa Fe River Canyon 113
Santa Fe Trail 25, 62, 104
Santa Rosa 62, 63
Santa Rosa Formation 35, 87, 103
Santo Domingo Basin 112, 114, 117
Santo Domingo Pueblo 60, 106, 113, 114, 116
Segundo Alto surface 121, 123, 163

Sierra Blanca 92
Sierra Ladrones 7, 19, 74, 99, 144, 150
Sierra Ladrones Formation 23, 145, 151, 154
Socorro 23, 27, 28, 29, 55, 144, 150
South Domingo Baca Arroyo 75
South Mountain 91, 102
South Valley 64, 143, 144
Spanish colonial rule 25, 33, 42, 44, 56, 58-61
Sulphur Springs Canyon Picnic Ground 88
Summerville Formation 86, 87, 156
Summit House 73, 91
Suwanee basalt flow 154, 156, 157

T
Taylor Ranch 55, 123
Tecolote 81
Tecolote Peak 91
Tejano Canyon 90
Tetilla Peak 112
Tiffany mine 101, 107
Tiguex Province 56
Tijeras 62, 79, 81, 83, 85, 87
Tijeras Arroyo 5, 80, 82, 144, 145
Tijeras Basin 86
Tijeras Canyon 17, 19, 26, 32, 42, 53, 55, 56, 77, 79, 80, 81, 83, 90
Tijeras Creek 5, 80, 81, 82, 87
Tijeras fault 82, 83, 84, 85, 86, 87, 102, 103, 105
Tijeras greenstone 17, 83
Tinkertown Museum 87
Todilto Formation 21, 86, 96, 105, 115, 118, 156, 158
Tomé Hill 23, 31, 148
Tonque Arroyo 62, 102, 117
Tonque Arroyo Member 156
Treaty of Guadalupe Hidalgo 45
Trigo Canyon 22
Tuerto gravel 103, 108, 111

turquoise 25, 33, 34, 101, 107, 109, 130, 151
Turquoise Trail 85, 87, 100-110

U
U.S. Route 66 43, 62-64, 82, 84, 117, 132, 142, 158
University of Albuquerque 135, 163
University of New Mexico v, 26, 29, 61, 62, 64, 78, 80, 84, 101, 127, 137, 140, 142, 143, 151

V
Valles caldera 92, 106
Vargas, Don Diego de 104, 129

W
Waldo 110, 111
Wind Mesa 150
Wind Mesa volcano 23, 31, 147, 148, 150

Z
Zamora 81
Zia Formation 22
Zia Pueblo 60, 116
Zuni Mountains 155
Zuni Plateau 26
Zuni Pueblo 55, 56, 60, 159
Zuni Sandstone 156, 159

PHOTOS & ILLUSTRATIONS

Individual photographers, artists, and archives hold copyrights to their works. All rights reserved.

Unless otherwise noted, all illustrations were produced by the staff of the New Mexico Bureau of Geology and Mineral Resources.

Albuquerque Museum 1 (#1971.061.001), 63 (top: #1990.013.002; bottom: 1982.180.211), 65 (top: 1980.075.13), 90 (#1980.061.001), back cover (#1990.013.002)

Pete Balleau 115

Paul Bauer 73, 154 (top), 155 (bottom), 161

Tom Bean 138, 159

Black Hills Institute of Geological Research 18 (bottom; illustrations by Dorothy Sigler Norton)

Ron Blakey 18, 20, 22

Center for Southwest Research, University of New Mexico 45 (top), 61 (top & bottom)

Earth Data Analysis Center 2 (top & bottom), 144 (top & bottom)

Jacqueline Guilbault 148

Adriel Heisey 38, 47, 49, 137

George H.H. Huey cover, i, viii, 9, 10, 12, 17, 51, 53, 55 (Awanyu bowl from Pecos Pueblo), 56, 59, 70, 92, 106 (bottom), 107 (bottom), 111 (bottom), 122, 123, 130, 158

Peggy Johnson 46

Paul G. Logsdon (courtesy of Marcia L. Logsdon) 105, 113, 150

Dave Love 118, 151 (top)

Rick Lozinsky 88

Spencer Lucas 19 (bottom)

Douglas Magnus 107 (center)

Marble Street Studio cover (inset), iii, x, 3 (top & bottom), 62, 120, 162, 168, 169 (top & bottom)

Maxwell Museum of Anthropology, University of New Mexico 101 (#60.24.17), 151 (bottom)

Middle Rio Grande Conservancy District 132 (#603, plate 1), 147 (top: # 353, plate xiv)

Museum of New Mexico 69 (#3371; photo by John K. Hillers), 108 (#9554), 109 (#9314), 116 (#49262), 124 (#118, plate 2), 126 (#57809), 127 (#48392), 129 (#100301).

Museum of Northern Arizona 57, 128

New Mexico Bureau of Geology and Mineral Resources photo archives 33, 34, 35, 103 (courtesy of Robert Shantz)

New Mexico Magazine 65 (bottom)

New Mexico Museum of Natural History and Science, courtesy of Spencer Lucas 84, 114, 145

New Mexico Office of the State Engineer, courtesy of Doug Heath 152, 153

New Mexico State Records Center & Archives 27 (Joseph Smith, photographer)

Greer Price vi, 45 (bottom), 76, 80, 83, 86, 89, 94, 97, 103, 106 (top), 110, 111 (top), 125, 133, 146, 149, 152 (top), 155 (top), 156 (both),

Gary Rasmussen 50

Adam Read 7, 75, 135, 154 (bottom)

Jeff Scovil 11

Joseph Edward Smith Photography Museum and Gallery 27

Bill Stone 8, 71, 74, 98

Amanda Summers 18-19 (bottom), 52-53 (bottom)

Kent Stout (sketches) on 94 (bottom), 160 (after H. M. Wormington and Arminta Neal)

Texas Tech 24-25 (courtesy of Bonnie Roach and Daniel Weissman)

University of Utah Press 16 (painting by Carel Brest van Kempen)

U.S. Army Corps of Engineers 147 (bottom: AD-32-593)

U.S. Geological Survey 139

Acknowledgments

We gratefully thank the many people who helped with this book. Charles E. Chapin, emeritus state geologist, and Peter A. Scholle, state geologist, unwaveringly supported this project. We are indebted to the New Mexico Geological Society, whose guidebooks (in particular 1961, 1982, 1995, and 1999) served as important references and provided some additional material. Peggy Johnson assisted with logging and provided valuable advice on content. Doug Rappuhn, Brian Honea, and Keith Miller assisted in unique and creative ways. Paul Brown, Kathy Glesener, Michiel Heynekamp, David McCraw, Rebecca Taylor, and Jan Thomas combined creativity, artistry, and skill in the production of many of the superb drawings early on. Glen Jones furnished invaluable assistance on GIS applications. Thanks to Imogene Hughes and Ed Smith for access to the Bonanza Creek Ranch and the Tiffany Mine. Robert Eveleth provided all information for historic photo captions.

We are especially grateful to Rick Aster, Charles Chapin, Sean Connell, Glenna Dean, Doug Earp, Barry Kues, David Love, Spencer Lucas, Gary Smith, and Frank Titus for taking the time to provide us with beneficial manuscript reviews. Spencer Lucas also provided written material for the chapter on geologic history.

We would also like to thank David Beining at the Explora Science Museum; David Brugge; Ann Carson and Mo Palmer at the Albuquerque Museum; the Center for Southwest Research at the University of New Mexico; Joseph R. Garcia and Cynthia Saiz of Our Lady of Sorrows Church in Bernalillo; Jose Guzman at Coronado State Monument; Louanna Haecker, Arthur Olivas, and Steve Townsend at the Museum of New Mexico; David Kammer; Joe C. McKinney at the University of New Mexico; Matthew Schmader at the City of Albuquerque; Alan Shalette; M. Kent Stout and Erik R. Stout, Quivira Research Center; and James Walker of The Archaeological Conservancy.

About the Authors

Paul Bauer is senior geologist and associate director at the New Mexico Bureau of Geology and Mineral Resources. He is manager of the Geologic Mapping Program, and an adjunct faculty in the Earth and Environmental Sciences Department at New Mexico Tech. He has published numerous books, papers, and geologic maps on a variety of topics of New Mexican geology. Transplanted from Massachusetts in 1979, he earned a Ph.D. in geology from New Mexico Tech in 1988. He and his wife, Peggy Johnson, live on a small farm along the Rio Grande in Socorro County.

Paul Bauer (right) and Rick Lozinsky.

Rick Lozinsky has been studying New Mexico's geology since 1979, specializing in basin-fill stratigraphy of the Rio Grande rift. His Ph.D. dissertation from New Mexico Tech was a pioneering investigation of the sedimentary deposits of the Albuquerque Basin. He has written several papers on the subject, and is senior author of the Elephant Butte Scenic Trip. When Rick is not teaching geology in southern California, he is usually trekking to some far corner of the globe.

Carol Condie has been involved in anthropological, historical, and cultural resource management projects in the Southwest for over 25 years. She holds a Ph.D. in anthropology from the University of New Mexico, and is president of Quivira Research Center and owner of Quivira Research Associates, firms that conduct archaeological, historical, and historic architectural investigations for private companies and for federal, state, and local agencies. She has three children and lives in Albuquerque with two three-legged cats, one four-legged cat, and a one-eyed blue heeler.

L. Greer Price is currently senior geologist and chief editor at the New Mexico Bureau of Geology and Mineral Resources, where he manages the publishing program. Prior to coming to the bureau, Greer spent four years as managing editor at Grand Canyon Association, ten years with the National Park Service, and eight years as a geologist in the oil patch. His varied career has involved teaching, writing, and field work throughout North America. He is the author of *An Introduction to Grand Canyon Geology*.

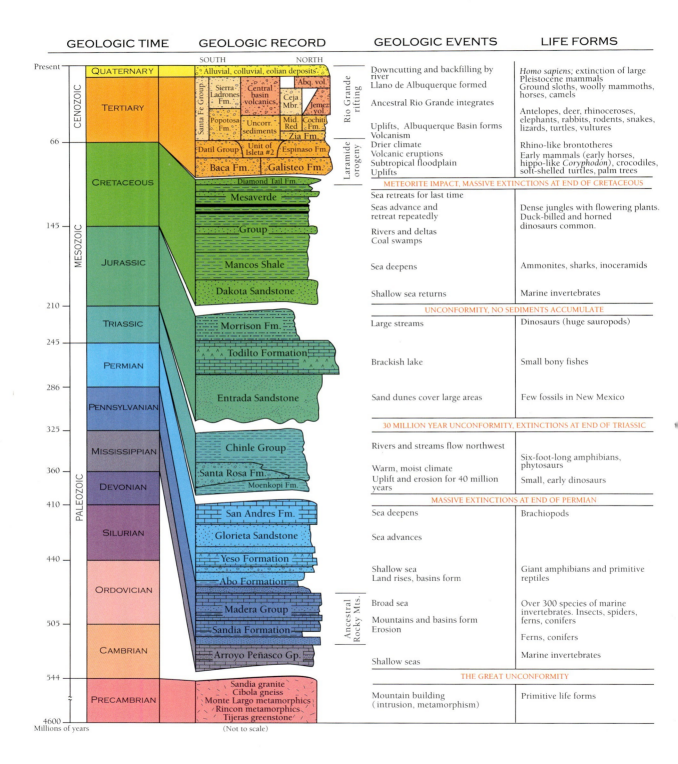